SURFACE WEATHER MAP AND SPECIMEN STATION MODEL

Published by the U. S. Department of Commerce - NOAA

Cloud type. (High cirrus.)

Total amount of clouds. (Sky completely covered.)

Wind speed. (18-22 knots.)

Direction of wind. (From the northwest.)

Temperature in degrees Fahrenheit.

Visibility. (3/4 mile.)

Present weather. (Continuous slight snow in flakes.)

Dewpoint in degrees Fahrenheit.

Cloud type. (Low fractostratus and/or fractocumulus.)

Height of cloud base. (300 - 599 feet.)

Part of sky covered by lowest cloud. (Seven or eight tenths.)

Cloud type. (Middle altocumulus.)

Barometric pressure at sea level. Initial 9 or 10 omitted. (1014.7 millibars.)

Amount of barometric change in past 3 hours. (In tenths of millibars.)

Barometric tendency in past 3 hours (Rising.)

Sign showing whether pressure is higher or lower than 3 hours ago.

Time precipitation began or ended. (Began 3 to 4 hours ago.)

Weather in past 6 hours. (Rain.)

Amount of precipitation in last 6 hours.

34 147

3/4 ✱✱ ● +28

32 - - - 6 ● 4

2 .45

Abridged from International Code

BOATING

How to Predict It,

WEATHER

What to Do About It

Sallie Townsend
Virginia Ericson

DAVID McKAY COMPANY, INC.
New York

Library of Congress Cataloging in Publication Data

Townsend, Sallie.
Boating weather.
Includes index.
1. Weather forecasting. 2. Meteorology, Maritime.
3. Boats and boating. I. Ericson, Virginia, joint
author. II. Title.
QC995.T67 551.6'5'1620247971 77-18015
ISBN 0-679-50798-1

10 9 8 7 6 5 4 3 2 1
Manufactured in the United States of America

CONTENTS

ILLUSTRATIONS

Figures 2–5, 7–16, and 19–21 by Sallie Townsend
Metric Conversion Scales on Figures 24, 27–29 by G. Apffel Pierce

ACKNOWLEDGMENTS

Our heartfelt thanks and deep appreciation are extended to the many persons who helped us amass an unbelievable amount of data relevant to weather during the five years it took us to write this book and to those specialists who gave analytical critiques of portions of the completed manuscript. (If we have missed anyone in these pages, please forgive us; our files have survived four complete moves between us during this time and a few letters and notations may not have made it.)

Mr. Max W. Mull, chief, Marine Weather Services Branch of the National Weather Service assisted us in gathering information as well as by reviewing the manuscript and writing the foreword for the finished book.

An early version of the general weather coverage of Part I was reviewed by Dr. Frederick Sanders, a professor of meteorology at Massachusetts Institute of Technology. Sandy is a veteran of many ocean races in his sloop *Stillwater* and is a personal friend of ours. The beginnings of Sallie's deep interest in the study can be traced to the time when he was her instructor for the U.S. Power Squadron Weather Course.

We are pleased that William S. Cox could take time from his sailing and sunning to review the manuscript while visiting the Townsends in Florida. Bill is widely recognized as an outstanding yachtsman, as is attested to by his being chosen in 1974 for the Nathanial G. Herreshoff Trophy, the highest honor the United States Yacht Racing Union can bestow.

The broadcast chapter was reviewed by Theodore Wyzewski, commander, NOAA; Vincent Kajunski, senior engineer, FCC, Boston, Mass.; William G. Konos, radio officer on thirteen merchant ships, radio officer for four years with the U.S. Coast Guard, and presently owner of a marine electronic equipment and supply business in Beverly, Mass.; Ramsey McDonald, electrical engineer with many patents in mobile radio communication; Virginia's husband, John, and Sallie's son, Sam, Jr., who number chemical and electrical engineering degrees and extensive experience in the field among their other accomplishments.

Hurricanes and waterspouts were initially researched and the material then examined through the courtesy of Dr. Neil Frank, director, National Hurricane Center, Miami, Fla.

Our authority on the stormy-weather chapter was Sallie's husband, Sam, by virtue of the fact that Sallie and Virginia consider him the best seaman they have ever known.

Part II of this book, which is the discussion of weather for most of the major boating areas of the United States, is the result of one of the many excellent suggestions given to us by our friend and agent, John Burton Brimer. The Great Lakes chapter of this section was reviewed by John Hanna, public affairs officer of the Great Lakes Region of NOAA; and C. R. Snider, meteorologist-in-charge (MIC), National Weather Service, Detroit, Mich.; the Atlantic Coast by Robert Lynde, marine forecaster, National Weather Service, Boston, Mass.; the Gulf Coast by Billy J. Crouch, MIC (acting), National Weather Service, New Orleans, La.; and the Pacific Coast by Robert A. Baum, National Weather Service Forecast Office, Redwood City, Calif.

Others in a variety of offices of the Department of Commerce's National Oceanic and Atmospheric Administration granted us interviews, responded to our queries by letter, or gave us pertinent written material. These include:

Thomas H. Reppert, Warren Hight, Bob Schoner, Jerry LaRue, John Galt, Al Peters, and James Travers at the Washington offices of the National Weather Service, as well as Dr. Duane Cooley, Chief, Technical Procedures, and Stanley Doore, the metrication specialist for this department;

A. I. Cooperman, chief, Data Information Group, Environmental Data Service;

Edwin Weigel and William O. West, Office of Public Affairs;

Edward Kohler, Environmental Data Service;

Frederick F. Ceely, Jr., chief; William Stanley, Scientific Data and Service; Captain Donald Tibbit, chief of Marine Chart Division of NOAA; and Jim Gearhart, staff cartographer; all with National Ocean Survey;

William H. Haggard, director, National Climatic Center;

R. H. Simpson, past director, and Arnold Sugg and Robert O. Cole, past and present deputy directors, National Hurricane Center;

Dr. Joseph H. Golden, National Severe Storms Laboratory;

Wilbur F. Mincey, supervisor, NWS, West Palm Beach, Fla.;

Wendell Porth, MIC; and Milton Williams, both of NWS, Tampa, Fla.; Buckley Christian of West Palm Beach, Fla.; H. D. Dyke of Cleveland, Ohio; and Edward M. Carlstead, MIC, Honolulu, Hawaii; all marine meteorologists.

In the U.S. Coast Guard, we especially want to thank Admiral A. C. Wagner, commander 7th District, Miami, Fla., who wrote the foreword for our book *The Amateur Navigator's Handbook* when he was chief, Office of Boating Safety, and who frequently assisted us in this book. Others in this branch of the services who should be singled out are Commander J. F. Smith, chief, Boating Safety in District 13, Seattle, Wash.; Lt. Commander Garth, Tele-Communications CB Expert, Washington, D.C.; Lt. R. I. Harrington, Galveston, Texas; Lt. Commander H. R. Little, chief, Audio and Visual Aids Branch, Washington, D.C.; Lt. L. M. Lilly, Member Training Branch; and Fritz Du Brucq, chief warrant officer, ret., who spent many hours with us imparting a great deal of information.

We would like to mention Mr. James Munson, Miami, and Mr. Gerald Sarno, Boston, of the FCC; George N. Weston, director, Navigational Sciences Division of the U.S. Navy; Chester H. Page, International System of Units coordinator, and Fred McGrehan, both at the Bureau of Standards, Washington, D.C. Many persons in the U.S. Navy headquarters in Boston and Palm Beach gave us considerable help as well as several in the Federal Aviation Administration in Boston and the National Bureau of Standards in Boulder, Col. We also received much useful material from the following districts of the U.S. Army Corps of Engineers: New England; New York, N.Y.; Philadelphia, Pa.; Norfolk, Va.; Jacksonville, Fla.; New Orleans, La.; Vicksburg, Miss.; St. Louis, Mo.; Chicago, Ill.; and Sacramento, Calif. To all those unseen hands that did not sign a name to their efforts, we are grateful.

Persons in the following Chambers of Commerce gave us assistance: Kitty Hawk, N.C.; Charleston, S.C.; Fernandino, Englewood, and Panama City, Fla.; Mobile, Ala.; Biloxi, Miss.; San Diego, Calif.; Seattle, Wash.; and Juneau, Alaska.

Others who assisted us in seeking information in Part I are: F. Dike Mason, instructor, U.S. Power Squadron; Fred Woods, nautical supply store owner in Marblehead, Mass.; Robert Snyder, Snyder Oceanography Services, Jupiter, Fla.; Tom Chamberlain, marine electronics store owner in Stuart, Fla.; Robert Grubb, communications store owner in West Palm Beach, Fla.; Frank Signaigo, an experienced electronics enthusiast; John Scott Herrmann, metrication authority; and Virginia's son Jim, a private pilot since age 16. G. Apffel Pierce, an analytical engineer, did the drafting for the metric conversion scales in the Appendix.

Between us, we have participated in pleasure boating in all the regions of Part II to varying extents but in many cases have also relied on others for descriptions of local weather conditions. We are exceedingly grateful to those who gave interviews, wrote informative lengthy letters, filled in extensive questionnaires, or tracked down knowledgeable persons. In addition to those mentioned in the text, we are appreciative of all the effort expended by the following persons:

Great Lakes: Desta and Bruce Bailey, Dean Baker, Charles Baker, Maureen and John Belka, Marjorie Brazier (author, *Wind off the Dock*), Susie and Donald Coulter, Lois and Walt DeVries, Dee and Jim Dilworth, Bud Doyle, Robert Hamilton, Ned Keys, Bev and Bud Marshall, John Norton, John Potter, Margery and Don Schantz, Meredith Scott, and Herman Vander Leek of the United States and Wilbur Bennett, Arnold Burns, Jeanne and John Wayling, and personnel from Information Canada of Canada.

Atlantic Coast: Bruce Brett, Pat Gerlach, Russ Lamont, Robert Searle, and Margaret and Whit Whittum.

Gulf Coast: Bill Burchfield, Ralph Damp, Neil Hutson, Belton Johnson, Hazel Knowlton, Arthur Nazro, Bobbie Strang, Bud Tritschler, Amy Waddington, and Brita and Roger Wooleyhan.

Pacific Coast: Vern Albert, Pat Brockhurst, Judy and Mike Crutcher, Bee and Hal Dearing, Cy Gillette, Florence and George Irving, Harry Monahan (editor, *Sea* and *Sea Guides* for the Pacific Coast), Jo and Dale Mogle (authors,

Sea's Hawaii Cruising Guide), Priscilla and Gar Morse, and Robert Walker, director, California Department of Harbors and Watercraft.

Thanks also go to our patient proofreading friends, Pauline Kurz and Barbara Harvey. Sallie's daughter, Lee, a physicist, deserves a particularly warm note of appreciation. She taught herself wave theory to explain tsunamis to us, read the entire book each time it progressed to another stage, and finally commented, "This is really the amateur weather-watcher's handbook; it tells you all the good stuff you only know if you've done it!"

FOREWORD

Boating Weather goes far toward taking the guesswork out of weather problems the mariner encounters. Although it is written for those who do not have any special training or experience in coastal or offshore weather, it contains information that will be useful even to those who do.

Many of the problems on the water come from not having a knowledge of, and preparing for, impending weather hazards. Certainly, every boatman should have some means for keeping up with the warnings and advisories issued by the National Weather Service. Chapter 5—"Utilizing Weather Broadcasts"—explains the many ways to do this. But, as the authors point out, just getting the forecast is not enough. The mariner should know that the storm described in the weather forecast may move faster or slower than predicted. Its path may change, and the local winds and waves may be affected by shore topography, channels, distance to shore, and other factors. The boatman who understands the relationships between clouds, winds, waves, visibility, and other weather elements is much less likely to find himself in trouble from "unexpected" weather. This book not only does a very good job of explaining weather systems generally, but also has a wealth of information on regional and local weather problems the mariner may encounter. In short, it is highly recommended for both the novice and the "old salt."

Max W. Mull
Chief, Marine Weather Services
National Weather Service, 1965–1977

WATERSPOUT ACTIVITY *A group of waterspouts at varying stages of development, photographed in waters of the Bahamas. An official photograph, NOAA.*

PART I

Investigating the Weather

The first part of this book concerns the information necessary for a realistic appraisal of weather conditions, past, present, and future, with emphasis on the application of this information to recreational boating in particular. The second part, Researching Regional Weather, deals with specific weather conditions likely to be encountered in the major boating areas of the United States.

Although National Weather Service personnel tell us that the adoption of the metric system is a sure thing, it is impossible at this writing to forecast the timing and extent of the changeover. In order to cover all eventualities, whenever metric measurement units are used in this book, they are followed by U.S. customary units, such as kilometers (miles), degrees Celsius (degrees Fahrenheit), and the like. Metric conversion tables pertinent to measurements of weather conditions appear in the Appendix. Since this book is concerned primarily with the effect of weather on boating, nautical miles and knots, the distance and speed measurements of the sea, are used whenever possible.

A new dimension is added to weather awareness when you become interested in boating. Most landlubbers think of weather changes as inconveniencing land-based recreational plans such as a day at the beach, a game of golf, or a backyard barbecue. It is highly unlikely that family or guests will be endangered if these plans are carried out despite adverse weather conditions.

On the water, however, you soon discover that the captain of any craft is responsible not only for the comfort but also for the safety of everyone aboard. Coping with worsening weather is not so simple as it miles per hour) and to travel about 800 kilometers (500 miles) in one day. This average speed tends to increase to 48 kilometers per hour (30 miles per hour) in the winter and thus the air masses travel about 1100 kilometers (700 miles) per day.

Most of the warm air masses that influence weather in the continental United States originate either in the southwestern states or in northern Mexico and tend to move in a northerly direction during their overall eastward progression. Cold air masses, on the other hand, usually originate in Canada and have a southerly trend as they progress across the United States toward the Atlantic coast.

Within the overall pattern of the summer and winter average rate of easterly movement, weather systems tend to maintain the same speed and to travel in the same direction as they did on the previous day. They are also likely to maintain approximately the same intensity.

For example, if you do your summer boating on Massachusetts Bay and hear of severe thunderstorms, gusty winds, and unseasonably cold weather in the Great Lakes region on a Thursday, you should expect this type of weather to arrive at the coast in about two days. As the storm's location on Thursday is to the northwest of you and the distance is approximately 1600 kilometers (1,000 miles) away, your prospects for a pleasant weekend on the bay are very gloomy indeed.

The National Weather Service records the progress of weather systems across the Pacific Ocean and North America on consecutive hourly weather maps. These are essential for their up-to-date reports as well as being important aids for their weather predictions.

WEATHER MAPS

You will find a comparison of sequential weather maps to be the real tip-off to approaching weather. Surface weather maps, especially those published by the National Weather Service, are a helpful basis for discussions of weather conditions and weather systems. Figure 1, which appears on both the front and back endpapers, is a reproduction of one of these maps. We suggest you refer to this map and its accompanying key while reading the rest of this chapter.

A weather map such as this is prepared daily at National Weather Service headquarters in Washington, D.C. from information received at 7:00 A.M. Eastern Standard Time. These maps are compiled into a weekly series entitled *Daily Weather Maps*.* They are based on large

---

* May be ordered from Public Documents Department, U.S. Government Printing Office, Washington, D.C. 20402.

operational weather maps that are plotted hourly. Surface weather information for the large operational maps is supplied periodically to the National Weather Service headquarters by approximately 1,000 United States weather reporting stations and 13,000 substations.

Material from the reporting stations is recorded each hour in code form on operational maps. Other symbols, drawn over the reporting station data, depict the weather systems present over each area at that time. Once the relationship of local weather conditions to large weather systems is understood, you will be able to interpret weather information on weather maps and from other sources discussed later in this book.

SPECIMEN STATION MODEL

A graphic explanation of the format for plotting station reports on daily surface weather maps is entitled Specimen Station Model. As you can see, this is also part of Figure 1 on the endpapers. This model appears in each weekly edition of the *Daily Weather Maps*.

Since we consider a study of this model an excellent way to become more familiar with meteorological terms and with related National Weather Service surface weather maps, we use it as a takeoff point in getting our weather discussion underway.

This Specimen Station Model was printed prior to the Metric Act of 1975 and uses the measurement units customarily used in the United States at that time. See the Appendix for conversion tables for units you may now be hearing about and using.

REPORTING STATION

In the Specimen Station Model, the location of the station reporting the weather is the center of a round circle forming a nucleus for the information in the station report.

Weathermen who interpret the weather story for the general public often have to adapt the National Weather Service code when they draw their maps in order to make the information more understandable to their audience and more practical for the space available to their presentation. For example, on television maps, the names of the cities are often omitted and the telecaster gives the names verbally when explaining the symbols he has printed at their locations. Newspaper maps sometimes show only overall weather systems.

CLOUD COVER

The amount of the reporting station's location circle that is filled in with black shading is the percentage of cloud cover at the station at the time of the report. If the circle is in outline form, clear weather is indicated.

At the upper left hand side of the model's key are the words "Cloud type" with a line to the symbol in the sample station report for this type of cloud formation. Note that the key further identifies this symbol as standing for "high cirrus." The information for the three different cloud types appears in different places in the key; high cloud and middle cloud formations are given above the location circle and the low cloud layer appears below the circle. There are at least 36 ways to indicate cloud types in the weather code, but most of them are combinations of the symbols used for the three basic cloud formations—cirrus, cumulus, and stratus. These three main symbols are:

1. A small hook at the end of a line for the feathery cirrus clouds
2. A semicircular cup or dome for the mounds of cumulus clouds
3. A straight line for the layered stratus clouds

WIND INDICATOR

The direction of the wind is depicted by an arrowlike symbol that flies with the wind. The point of the arrow is embedded in the location circle. The speed of the wind in knots is shown by "feathers" at the end of the shaft:

One short or half-feather = 5 knots

One long or full feather = 10 knots

One triangular flag or pennant as a feather = 50 knots

Sometimes the wind indicator symbol will have a plus sign and a number adjacent to it, such as +30. The plus sign indicates gust winds and the number shows the peak velocity of the gusts. Wind speeds may be given either in knots for wind over water and sometimes over land, or in kilometers per hour (statute miles per hour) for wind over land; knots, generally used in this book, are nautical miles per hour.

TEMPERATURE

The air temperature readings on *Daily Weather Maps* are taken at 7:00 A.M. Eastern Standard Time as noted on your May 13, 1974 surface weather map; the temperatures are given in degrees Fahrenheit. Average temperatures for the day or an estimate of future temperatures may appear on other types of maps and may be given in degrees Celsius. Television weathercasters will often mark the low temperatures for the day on their weather maps during the winter months and the high readings in the summertime in order to add a little drama to their presentation.

An advancing air mass is primarily identified by its temperature in comparison to the temperature above the surface of the earth directly ahead of it. When this air is warmer than the temperature of the region into which it is moving, it is referred to as a warm air mass and, conversely, when the advancing air is colder, it is called a cold air mass.

VISIBILITY

Although the distance figure for visibility is given in statute miles and fractions of miles on the station model, you may be more likely to see or hear this figure given in kilometers. Visibility reported at a distance of more than 16 kilometers (9 nautical or 10 statute miles) is omitted on the map.

PRESENT WEATHER

In the key, the term "Present weather" refers primarily to precipitation and the symbols describe weather conditions reported at the station such as "continuous slight snow in flakes" in the sample.

We list the present weather symbols you are most likely to see used on weather maps. We include the winter symbols not only for those in warmer climes who enjoy recreational boating all year, but also for those who live in colder climates so that they can continue their study of weather maps during the winter months. Sometimes these symbols are used in combination.

Short horizontal lines in a column: fog, smoke, or haze. Two lines indicate light fog and three lines, heavy fog. Broken lines are used for patchy fog.

Comma: drizzle. Severity from intermittent to thick indicated by one to four commas.

Dot: rain. Severity from intermittent to heavy indicated by one to four dots.

Asterisk: snow. Severity from intermittent to heavy indicated by one to four asterisks.

Triangle with one side on the bottom and one point up: hail.

Triangle with one side on top and one point down: showers.

Capital letter R with arrowhead at lower right end: thunderstorm.

Letters such as F for fog, R for rain, and S for snow are often used for these elements on media weather maps. Sometimes, however, television weathercasters use the National Weather Service symbols on their maps. It will be a great source of satisfaction when you find yourself identifying these symbols before they are explained or time runs out on the broadcast.

DEW POINT

The dew point figure is the temperature below which the air can no longer hold the total amount of water vapor present in it. Since cool air can contain less water vapor than warm air, a drop in air temperature can result in the condensation of the excess moisture. This condensation is likely to occur when the air temperature drops to within 2°C (4°F) of the dew point temperature. The moisture in the air which was formerly invisible in its water vapor state may become visible in the form of dew, fog, clouds, rain, or snow.

BAROMETRIC AIR PRESSURE

Our planet is tightly wrapped in a heavy blanket of air referred to as the atmosphere. High above us this air is very thin, but it gradually compresses and becomes heavier as it nears the surface of the earth. The measurement of this atmospheric pressure, with an instrument called a barometer, is referred to as barometric air pressure.

All atmospheric air pressures given at weather stations are reduced to a sea level figure in order to have a standard for comparison. The barometric pressure figure on the reporting station key is the number in the upper right-hand corner.

Barometric pressure at sea level: this is given using millibars (mb) for the measurement. In the key, the last three digits of a millibar measurement are used. Therefore, the actual reading of 1014.7 mb is given as 147.

Millibars are the units of measurement used by the National Weather Service on the surface weather map (Fig. 1), although many weather reports in newspapers or broadcasts on television or radio use inches of mercury (in) for the measurement or kiloPascals (kPa), the metric units adopted to measure atmospheric air pressure. Conversion tables appear in the Appendix.

Although a noticeable high pressure is usually associated with good weather and a marked low pressure with inclement weather, the important clue to forecasting is the trend of barometric air pressure during the previous three hours. Generally, a rising barometer foretells the weather is clearing; a falling barometer, the weather is worsening; and a steady barometer, the conditions are static.

Directly under the barometric pressure at sea level figure on the key are the symbols that indicate the pressure trend:

Amount of barometric change in past 3 hours: this is given in millibars in the station model with the decimal point moved one digit to the right. The figure of 28 on the model represents 2.8 mb. When there is a change, the number is preceded either by a plus or minus sign:
+ sign = pressure is higher
− sign = pressure is lower

Barometric tendency in past 3 hours: one short line in the station report describes the barometric trend during the previous 3 hours.
Slash mark = rising barometer
Reverse slash mark = falling barometer
Horizontal line = steady barometer
Wiggly line = unsteady barometer

A second line is added to the end of the first if the tendency has changed during the 3-hour period. For example, when a slash mark has a horizontal line extending from the top of it, the symbol indicates a barometer that rose initially and then became steady.

WEATHER IN PAST 6 HOURS

The symbols used to show weather conditions during the past 6 hours are the same as those used to describe present weather conditions, such as a comma for drizzle or a dot for rain, and are followed by the time the weather conditions began or ended.

The amount of any precipitation occurring in the last 6 hours is given in inches in the specimen station model. Centimeters are the metric units used to measure precipitation.

WEATHER SYSTEMS ON WEATHER MAPS

The location of large weather systems is determined from a comparison of the coded station reports on weather maps. Symbols illustrating the presence of these large systems are then superimposed over the data recorded at the location of each station. The resulting graphic presentation makes it possible to see at a glance the location of weather masses close to the surface of a particular area. In the overall weather picture of the continental United States, as you know, the large weather masses with different air temperatures and varying barometric air pressures progress across the country in the general direction of west to east, with warm air moving up from the south and cool air down from the north.

HIGHS AND LOWS

A region of high barometric air pressure is called a high pressure area or a high and is determined by a comparison of the barometric pressure readings from reporting stations. The highest barometric pressure in an air mass is in the center of the high and is identified on National Weather Service maps by the word HIGH or the letter H. A low pressure area, or low, which is a region of comparatively lower barometric pressure, is shown on their weather maps by the word LOW or the letter L.

Clear weather is generally found in and ahead of high pressure areas. Surely, a recreational boatman's idea of heaven includes a local high barometric pressure area every weekend during the boating season. Weather-wise boaters, however, direct most of their attention to keeping track of the low pressure areas, which are the villains in the weather drama as they are usually associated with bad weather.

ISOBARS

On surface weather maps, irregular lines encircle or partially surround high and low pressure areas. These lines are called isobars.

Isobars are drawn to connect the locations where the same barometric air pressure was registered during the same reporting period and, thus, the barometric pressure is the same at any place along an isobaric line. Although barometric air pressure is not a solid substance such as land, the analogy can be used to consider its highs as elevations like high ground or mountains, and its lows as depressions like lowlands or valleys. Therefore, isobaric lines are like contour lines drawn on topographic maps. Furthermore, sea level is the datum used for both measurements, barometric air pressure and land elevations.

On a worldwide weather map, each isobaric line is completely closed, as it encompasses either a high or low pressure area. However, even on a map portraying an area as extensive as the continental United States, the portion of the earth's surface represented is generally too small to show many closed isobars. Of necessity, therefore, many of these lines end at the borders of the maps.

Isobars on your National Weather Service *Daily Weather Map* are drawn for every 4.0 mb of barometric pressure and each line is identified with its measurement. On this surface weather map, the measurement figure on an isobar is a whole number and the tenths are omitted. On the Specimen Station Model, however, the same information is handled differently, as the initial digits 9 or 10 are omitted and the tenths of the measurement are included. For example, the barometric pressure at Houston, Texas, on your map is recorded as 128 at the station, which stands for 1012.8 mb of pressure, whereas the isobar drawn nearest this area is marked 1012.

Newspaper maps usually show fewer isobars than are drawn on National Weather Service maps. The measurements may be in inches of mercury, millibars, or the metric unit, kiloPascals. The legend accompanying the map will identify the unit used.

WIND CONDITIONS

Nature continually strives to maintain equilibrium in the atmosphere. This is demonstrated by the tendency of air to flow from higher pressure areas to lower pressure areas. This flow of air close to the earth is referred to as surface wind. In the Northern Hemisphere, surface winds flow in a clockwise spiral out from the centers of highs and then gradually change direction to spiral counterclockwise in toward the centers of lows. (Air spirals in exactly the reverse way in the Southern Hemisphere.)

Note how the direction of the wind on your surface weather map usually follows this general pattern as indicated by the arrows flying toward the location circles of the reporting stations.

Winds increase where there is a marked difference in the barometric pressure. A rapid barometric change, or steep pressure gradient, is represented on a weather map by a tight concentration of isobars. This is graphically illustrated by Fig. 17 on page 142, which is a map of a typical northeaster, the type of severe storm prevalent along the New England coast. Since this map appeared originally in a pamphlet on northeasters compiled some time ago by a government weather agency, the isobars are drawn for every 4.0 mb. A comparison of the closeness of the isobars around the "x" at the center of the low on this map with the isobars around the center of lows on your National Weather Service map dramatically demonstrates the steepness of the pressure gradient in a rigorous weather system.

Note on the northeaster map how strong the winds are at the surface reporting stations: it is blowing 45 knots (4 long feathers and 1 short feather) to the north of the center inside the 976 mb isobar and 60 knots (1 pennant and 1 long feather) southeast of the center inside the 1004 mb isobar line. Heavy rain is falling at the station to the east of the center inside the 984 mb isobar as indicated by the four dots. (At this reporting station, the pressure is recorded as 83. As there are no isobars under 1000 mb on the weather map on the endpapers, this is an example of the way a barometric pressure reading of 983 mb is written on a weather map by the National Weather Service.)

When you see a tight concentration of isobars drawn on a weather map to the west of your location, you can determine the approximate arrival time of strong winds in your area and make your boating plans accordingly. While a rapid change in barometric pressure denotes wind, widely separated isobars indicate light winds or none at all.

FRONTS

Low pressure systems are often associated with fronts. Fronts are lines of separation between cold and warm air masses that follow or overtake each other over the earth's surface. They are found generally in troughs of lowest pressure in an area. The barometric pressure decreases as it falls on one side of the front and increases as it rises on the other side of the front. Lines are drawn on weather maps to show the location of fronts at the earth's surface. You can assume that a front is likely to progress in a direction perpendicular to the frontal line, as shown on a weather map.

Fronts influence weather conditions on both sides of their surface locations because aloft their boundaries are never vertical. Instead, they slant either ahead of or back over the air mass that follows them with the warmer air overlying colder air. A front immediately precedes the air mass for which it is named.

Warm Front On National Weather Service maps, a line with half moons on one side indicates a warm front. The side of the frontal line on which the half moons are drawn shows the direction the front is moving. On some other weather maps, the words WARM FRONT are used along the frontal line instead of the half-moon symbols.

As a warm front approaches an area of cold air, the warmer air slips up and over the cold air. The upper part of the sloping warm front may be hundreds of kilometers or miles ahead of its location at the earth's surface (see Fig. 2, page 20).

As the warm air rises and cools, condensation occurs that can cause long periods of precipitation before the front passes by at ground level. This precipitation usually begins with brief showers and continues with from 6 to 24 hours of light, steady rain or drizzle and is frequently accompanied by fog. A decrease in barometric pressure, which can be very gradual, usually occurs with the approach of these weather conditions.

Persons venturing out on the water when a warm front is on the way should be prepared to navigate under conditions of poor visibility.

After the warm front passes, the rain stops, the barometric pressure steadies, and the temperature rises. There still may be some drizzle for a short time from low-lying clouds. Boaters should now be alert to the possible arrival of a cold front, which often follows closely behind a warm front.

Cold Front The cold-front line on our map has sharp pointed teeth projecting toward the intended path of the front. The words COLD FRONT are often used instead on the frontal lines on other weather maps.

A cold front moves faster than a warm front and wedges itself under the warm air mass ahead. It causes a marked change in the weather within a period of a few hours. The barometric pressure may fall rapidly.

The warm air is forced abruptly upward along the slope of the invading cold front. This abrupt action can produce turbulence, high gusty winds, and thunderstorms along and preceding the leading edge of the front (see Fig. 3, page 21).

The severity of the weather conditions depends on the rate of speed of the cold front. A cold front moving at a high rate of speed will force the warm air ahead to rise almost vertically. The moisture in the nses rapidly as the rising warm air is cooled and the pitation and turbulence develop into severe storms. A system in a deep low pressure trough is likely to produce efore and after the passage of the frontal system. When a noves more slowly, the cloudiness and precipitation are spreac r a greater area, causing the storms to be less severe, and, as the pressure gradient is not as steep, the winds will not be as strong. Usually, cold-front weather conditions persist no more than four hours.

Thunderstorms sometimes occur about 240 kilometers (130 nautical or 150 statute miles) ahead of a fast-moving cold front. A line of prefrontal thunderstorms are called line squalls. When there have been line squalls, thunderstorms almost never occur within the front itself.

As a cold front moves through any area, the wind shifts abruptly to the north or northwest. Usually, there is a noticeable drop in air temperature and sometimes the barometric pressure rises rapidly. Cold-front changes such as these are clearly evident in a study of station reports on your surface weather map. They are the primary signs a meteorologist uses to locate the cold fronts. Many feathers on the wind symbols at stations near a frontal line and a sharp drop in temperature at stations behind the line after the wind goes to the north are indications of a severe cold front. Once a cold front has passed, clear weather with diminishing clouds is likely to occur.

Boating plans should be reconsidered carefully when you expect the arrival of a strong frontal system in your area. Experienced yachtsmen usually elect not to go out at all at this time because of the probability of very high winds and seas. If, on the other hand, an expected frontal system is not likely to be severe, cruising plans can be arranged to take the best possible advantage of the weather changes that are likely to occur within a particular period of time.

For example, if a skipper of a sailboat expects the arrival of a mild cold front on Saturday night of his anticipated weekend cruise, he will have the wind with him both days if he plans to go to a harbor to the north of his home port. He will enjoy sailing before the south or southwest winds that precede the front while en route to his anchorage. The front with its possible disturbances should pass through that night. He will then have a comfortable return trip to his home port on Sunday, moving along with the wind again at his back.

Occluded Front Alternating half moons and sharp teeth on the same side of the frontal line depict occluded fronts on National Weather Service maps.

An occluded front occurs when either a cold front overtakes and merges with a warm front or the warm front is trapped between two cold fronts and the warm air is forced upward. The deepest low and, therefore, the most severe weather associated with frontal systems usually results from the meeting of a warm front and a cold front and is likely to occur within a 160-kilometer (90 nautical or 100 statute mile) area slightly to the north and ahead of the occlusion of the two fronts.

Stationary Front National Weather Service symbols for a stationary front show warm-front half moons on one side of a frontal line alternating in placement with cold-front teeth on the other side. (Your map does not include a sample of this type of front.)

When a front has stopped moving temporarily or is moving only slightly, it is called a stationary front. It is similar to a warm front in that it can bring long periods of precipitation.
the position of a weather system to the west of you, and analyze its

SHADED AREAS

Shading is used on National Weather Service maps for the locations where stations report some type of precipitation. It is possible, therefore, to see at a glance all the precipitation areas in the United States. When several neighboring stations report this weather condition, large areas are shaded. As might be expected, shaded areas are usually found near fronts. However, there may also be pockets of precipitation that are not associated with fronts, such as those shown near the Great Lakes on your weather map.

Some weathercasters select different types of shading to denote different kinds of weather conditions. If the words snow, rain, shower, and the like are not superimposed on a many-shaded map, there will be a key to the shading in newspaper maps or the weathercaster will explain their meaning on television maps.

NEWSPAPER WEATHER MAPS

Although newspaper weather maps contain less information than that found on the National Weather Service map you have been working with, they will resemble it in some ways. Often they are accompanied by a synopsis of weather conditions portrayed, as well as a key to the symbols and metric measurements used.

A map giving reported conditions at press time appears in some newspapers. Although this map is necessarily several hours old by the time the newspaper reaches most of its readers, its value lies in the on of actual weather conditions that have occurred rather prediction of things to come. On the other hand, many newsrint forecast maps based on predictions rather than actual cond ns.

If your newspaper prints an actual-condition weather map, you will be surprised at how much you will learn about weather systems when you save and compare daily newspaper maps within a period of a week or more.

TELEVISION WEATHER MAPS

Television weather broadcasts are usually more informative than newspaper presentations because they contain up-to-the-minute reported weather conditions. Of course, it is not possible to save the maps shown for study, reference, and comparison. Often the information telecasters write on their maps consists of weather reports received from the National Weather Service just prior to the broadcasts. Maps used on daily television weather broadcasts are based on National Weather Service surface weather maps and the same types of symbols and abbreviations are likely to be used.

The data given usually includes a résumé of conditions and forecasts on a country-wide as well as a local scale. Forecasting by means of this medium is unique because it gives meteorologists the opportunity to explain the reasoning behind their forecasts, the probabilities involved for the forecasts to occur as predicted, and the conditions that might cause them to change.

Satellite pictures of cloud coverage of the country and nearby ocean waters are often included, and sometimes a synchronized series of satellite pictures is used showing the progression of weather systems indicated by cloud movement. In satellite photographs, high pressure areas stand out because of the lack of clouds in them, fronts because of their long, thick cloud formations, and tropical storms or hurricanes because of the clouds that spiral around their central cores.

Weather reporting on television is simple and straightforward, as the information is primarily intended for a land-based layman. Often both metric and U.S. customary units are given.

TRY YOUR HAND AT FORECASTING

Although the mysterious forces of nature can be depended upon to refute at one time or another any sweeping statements about systematic weather performance, weather information recorded on maps coupled with expected weather movements make it possible to forecast local conditions with some degree of accuracy.

A comparison of consecutive weather maps will indicate the speed of advance and direction of movement of a weather system. However, when you know the time and date of a single weather map, a forecast also can be made, since the daily weather system's movement from west to east is about 800 kilometers (500 statute miles) in the summer and 1100 kilometers (700 statute miles) in the winter. The National Weather Service map you have been using includes a mileage scale for various latitudes and you can use it to make a rough estimate of the distance a weather system is from your boating area. The probable path of a front can be deduced from its position on a map because fronts are likely to continue in a direction perpendicular to their line and, in general, to maintain approximately the same intensity.

Take a look at today's newspaper or television weather map, note the position of a weather system to the west of you, and analyze its severity if possible. Calculate when it will probably arrive in your area. Follow its progress on later maps to note any changes and see how close its arrival time and strength are to your original predictions.

In addition to anticipating the arrival of large weather systems, you must also stay on the alert for conditions such as fog, wind, and thunderstorms that can develop locally. Nature's drama should command your close attention at all times as you examine the visual signs developing around you, detect weather changes with instruments, utilize as many weather broadcasts as you can, and probe into the cause and effect of nature's mighty whirlwinds to be sure a hurricane or a waterspout is not lurking on the horizon.

Did you spot the clues in the previous sentence and recognize them as the other chapters in Part I of this book? The mystery is just beginning to unfold. Understandably, the weather map is only the first step to the time when you can don your twin-billed master detective deerstalker cap with the gold anchor on it, nod sagely as the smoke from the strong tobacco in your briar pipe fills the air, and give forth your own elementary observations and weather predictions.

CHAPTER 2

Examine the Visual Signs

In ancient days, the uncanny power mariners possessed for predicting weather conditions was considered by many to be magical, bestowed on the deserving by the gods of the sea. Their reputations were based on their ability to recognize and understand the meaning of nature's earliest clues to changing weather patterns and to deduce from these clues the proper actions to take to safeguard their ships and all aboard. This weather awareness came from their recognition that one thing leads to another in Mother Nature's continual effort to maintain balance in her realm.

Even with the wealth of scientific weather data available now, mariners still employ the time-honored art of scanning wind, wave, water, sky, and horizon for weather signs. Any signs they discover are used to make today's extensive meteorological information more relevant to their local boating areas.

CLOUD FORMATIONS

Clouds are a visible sign of moisture in the atmosphere. Their formations are important clues to approaching weather conditions, although by themselves they are not always enough for reliable forecasting. However, since all you have to do to find these clues is to look up at the sky, learning to recognize their differences with the probable significance of each formation is a good place to begin an examination of nature's visual signs.

The three basic formations of clouds are the plumelike cirrus whisks, the puffy cumulus mounds, and the widespread stratus layers.

In translation from their Latin roots, cirrus is a curl of hair or fringe, cumulus is a heap or pile, and stratus is a coverlet or spread. Combinations of these basic cloud names are used to describe other kinds of cloud formations. Sometimes the words nimbus, meaning rain or storm cloud, and alto, meaning high, are also used.

A knowledge of the certain cloud formations that are usually associated with the approach of cold or warm fronts is particularly useful to boaters. These may be heralds of long periods of rain or poor visibility if a warm front is nearing, or possible turbulent weather conditions if a cold front is on the way.

The following are the cloud formations associated with warm fronts and cold fronts and the sequences in which they can occur. (Usually, however, only part of a sequence actually takes place.)

Warm Front Clouds

Cirrus. These feathery wisps of clouds are sometimes referred to as "mares' tails." Although they are considered fair weather clouds, they are often the first warning of an approaching warm front. They can appear as much as 1450 kilometers (800 nautical or 900 statute miles) ahead of the front and indicate that the front will probably arrive 36 hours later.

Cirrostratus. As the warm front approaches to within 1300 kilometers (700 nautical or 800 statute miles), this thin white veil of clouds can form over the sky and a halo appears around the dimmed outline of either the sun or the moon. When you spot this formation, you can expect the warm front to reach your location in about 32 hours.

Altostratus. The cloud cover thickens when the front approaches to within about 650 kilometers (350 nautical or 400 statute miles) or 16 hours before its arrival. The shape of the sun or moon may be obscured by the clouds although you will still be able to see a luminous light given off by them.

Stratus. This thick, low cloud layer is usually no higher than 300 meters (1,000 feet) above the earth's surface and may occur 15 to 16 hours ahead of and remain until the arrival of the warm front. The last vestige of the sun or the moon is usually completely blocked out, daytime hours are gloomy and nights are pitch black. Fog often accompanies this overcast. A stratus cloud formation can cover an area from the leading edge of a front to 650 kilometers (350 nautical or 400 statute miles) ahead.

Nimbostratus. This is a variation or development of stratus cloud layer. It is easy to identify because of its dark appearance and its tendency to produce a steady rain.

cumulus. After a warm front has passed through your loca- e brightening sky may be invaded by masses of these cotton- ouds riding high in the atmosphere. You will see them either as vidual tufts or as clusters.

Cumulus. Sometimes the clue to the passage of a warm front will be the appearance of a sharply defined billowing cumulus cloud formation instead of the tufted altocumulus formation. No matter which formation arrives on the scene, these fair weather clouds are always a welcome sight to mariner and landsman alike.

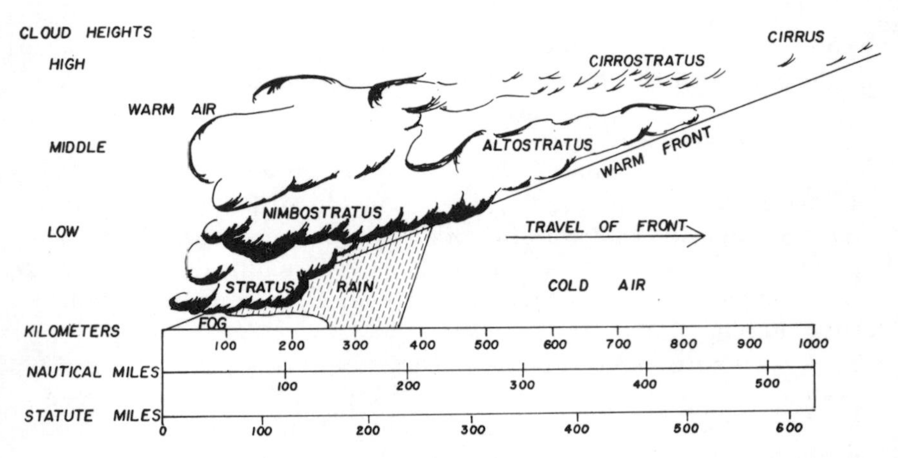

Fig. 2. WARM FRONT CLOUDS *Examples of cloud formations that may occur in connection with a warm front.*

Cold Front Clouds

Cumulus or Altocumulus. When you see the cumulus or altocumulus cloud formation, which formed after the passage of a warm front, begin to thicken, you can deduce that a cold front is on the way.

Cirrocumulus. Cirrocumulus clouds are cumulus clouds borne high in the sky. If you spot them thickening and descending from their lofty location, you can expect the approach of cold front storms in the near future. The rippled appearance of the clouds in this infrequent formation has earned it the nickname of "mackerel sky."

Cumulonimbus. When cumulus clouds darken to become ominously black, the storm-cloud appellation of nimbus is added to their name. They have become the familiar thunderclouds that unmistakably threaten severe rain squalls and gusty winds, including the possibility of hail. This awesome formation can tower up into the sky as

high as eight kilometers (4 nautical or 5 statute miles
face of the sea or land. Cumulonimbus clouds often im
cede the leading edge of a cold front.

Nimbostratus. Frequently, this low, dark cloud l
heavy rainfall also precedes a cold front. Although you c
the towering cumulonimbus formation will be above this . . . when
the nimbostratus layer forms, the cold front is probably about 160 kilometers (90 nautical or 100 statute miles) away. The heavy rainfall associated with this formation usually continues after the arrival of the front, sometimes for as long as a few hours after the front has passed.

Altostratus. When the nimbostratus cloud layer moves away from your location, you can assume that the cold front has moved on too. As this formation moves on, the sky may clear enough so that you can see a thick layer of altostratus clouds high overhead, which usually last only for a short period of time.

Cumulus. Once the upper part of the cold front has passed, scattered cumulus clouds may appear in the clear blue sky, a welcome sight after the turbulent cold-front weather.

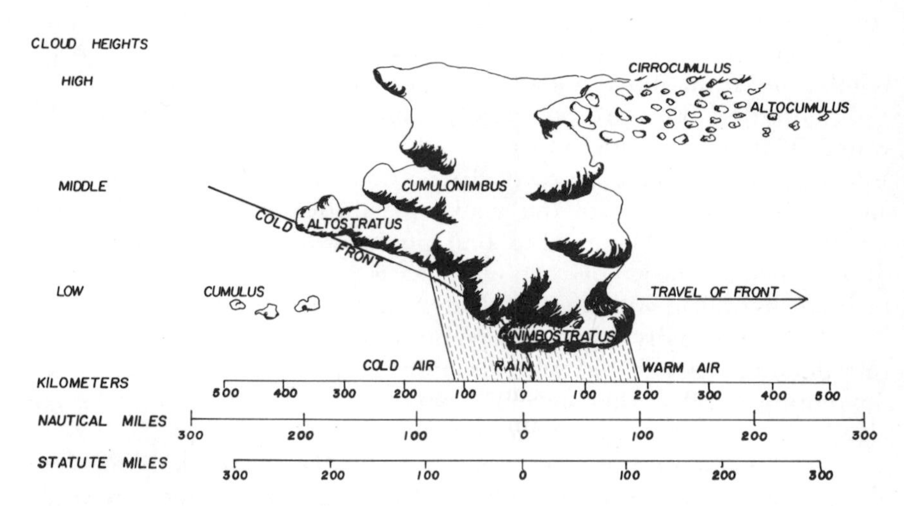

Fig. 3. COLD FRONT CLOUDS *Examples of cloud formations that may occur in connection with a cold front.*

Thunderheads Thunderheads are towering cumulonimbus cloud formations in which thunder and lightning storms are nurtured. They develop from a marked upward movement of warm moist air, which occurs, for example, when a fast-moving cold front wedges its way into a warm air mass and abruptly forces the warm air aloft. As you know, when warm moist air rises above the earth's surface, it cools and the moisture condenses. Sustained and forceful updrafts of warm moist air will swirl into swelling mounds of cumulus clouds, which are likely to darken and become a towering, ominous-looking cumulonimbus formation. The darkening of the cumulus clouds is an early sign to the possibility of thunderstorms. Once the cumulonimbus formation develops, the warning is unmistakable. According to a report from NOAA, "The transition from a small cloud to a turbulent, electrified giant can occur in as little as 30 minutes."

A thunderstorm is considered mature when the condensation in the cloud formation forms into liquid droplets that in turn increase enough in size and quantity so that they fall to earth within a pronounced down draft. Sometimes, as the storm reaches maturity, strong updrafts swoop up some of the droplets in the atmosphere where they are frozen into ice balls called hail. Hail can be tossed up and down within a turbulent thunderhead so that it grows in size in the process; it is coated with liquid each time it falls and when it is bounced upward again this new layer of water freezes. Hail falling to earth strikes surface objects with a resounding clatter. Even today, when we understand how the phenomenon occurs, we are amazed when we are bombarded from above with balls of ice during warm weather.

Thunderstorms are potentially dangerous to persons out on the water. The storms usually arrive suddenly with strong gusting and shifting winds, which in practically no time at all churn up the water and increase the size of the waves. Visibility is likely to become nonexistent while the rain or hail pours down in blinding, wind-driven sheets. The whole extravaganza is punctuated with flashes of brilliant lightning and claps of deafening thunder.

You can usually rely on the National Weather Service to report on thunderstorms associated with a rapidly moving cold front. However, developing local storm conditions sometimes make the recognition of visual signs in boating areas imperative.

In the summertime, afternoon thunderstorms are likely to occur along a coast when the humidity and temperatures ashore are high. This is particularly true along the Gulf and Atlantic coasts of the United States and along the shores of the Great Lakes. Heated air radiates upward from surfaces of the land that are exposed to strong rays of the sun. Moisture from a nearby body of water is absorbed by the warm air. As the warm moist air rises within the strong updrafts, the moisture begins to condense as the air cools, and swirling cumulus clouds form, which can ultimately become the black-looking cumulonimbus clouds. The cumulonimbus formation is often accompanied by the sound and light presentation of rumbling thunder and flashing lightning.

LIGHTNING

A thunderstorm is often referred to as an electrical storm because of the tremendous electrical charges that build up between the thunderstorm clouds and between the clouds and the surface of the earth below. When enough voltage has been generated within a storm cloud the giant electrical sparks, which we know as lightning, flash in the sky or streak down from a storm cloud to the surface of the earth where the electrical charge they carry is dissipated.

Lightning offers a particular threat to boats, projecting as they do above the relatively flat surface of the water, because of its tendency to be attracted toward the highest object in an area as it darts along its downward path.

One way to diminish the damage to a boat caused by a nearby or actual lightning strike, and to help protect the persons aboard, is to have the highest point on the boat connected with a large wire to a ground plate so that if the boat is struck by lightning the charge will follow the least harmful path to the water. There is a well-accepted theory that when the high point on a boat is grounded it tends to bleed off the charge to form a protective cone around the boat; the higher the high point the greater the area protected by the cone at the surface of the water will be.

As the mast on a sailboat is not only the highest point on the boat, but is also supported by metal shrouds and stays, adequate grounding is relatively easy to accomplish. It can be done in a number of ways such as running a grounding cable to all chain plates where the shrouds and stays are attached and connecting this cable to the keel bolts or to a special ground plate attached to the keel, or by simply attaching one end of a chain to a metal shroud and dangling the other end overboard about one-half meter (one to two feet) into the water. Aboard a power boat, the highest point on the boat should be grounded also. This highest point can be an actual lightning rod, flag mast, fishing outrigger, or the like. It should be metal, have as sharp a point as possible, and be connected to a ground plate.

In the event of a lightning strike, the electrical charge is likely to follow the path of least resistance provided by a good grounding system to the water instead of through electrical equipment, hulls, decks, or persons, with more disastrous results. You may wish to check with the manufacturer of your boat to see what type of grounding system, if any, was used when your boat was built.

A boat does not actually have to be struck by lightning for damage to occur to electrical and electronic equipment aboard; it may merely be in an area of storm-produced electrical disturbance. Lightning discharges produce charged particles in the surrounding atmosphere, and some transistors in electronic equipment are very susceptible to damage from these particles even though the voltage may be relatively low.

To guard against the possibility of future costly repairs, you might consider not only turning off all electrical equipment prior to a storm, but disconnecting it as well from the boat's electrical system. A radiotelephone should be disconnected from its antenna even though the antenna is properly grounded. The Ericsons learned by bitter and costly experience after a nearby lightning strike in Nantucket Harbor, Massachusetts, that simply turning off the radio receiver will not safeguard the set.

Anyone aboard who touches metal objects during a thunder and lightning storm may receive a shock. Since a shock, depending on its severity, can cause injury or even death, everyone should be warned of this danger. According to data assembled by NOAA, the average death toll caused by lightning in the United States is greater than that caused by tornadoes or hurricanes. Out of the 400 reported lightning accidents in one year, there were 150 deaths. Most of these could have been averted if the proper precautions had been observed.

As a person injured by a lightning charge may be shocked severely enough to stop breathing, mouth to mouth resuscitation should be administered as quickly as possible. There is no danger to the person giving first aid as the victim's body does not continue to harbor the electric charge.

Thunder is the sound wave produced when the heat generated by lightning causes a sudden expansion of the air. When you see lightning and hear thunder, it is possible to determine in nautical miles approximately how far away the thunderstorm is from your location. Count the seconds between a flash of lightning and the next clap of thunder and divide this elapsed time by 5.5. It takes sound about 5.5 seconds in average conditions to travel one nautical mile, and the speed of light is so great that the origin and the observance of the lightning in the sky can be considered to have occurred simultaneously. Thus, you "see" the thunder start and hear it arrive.

For example, if 22 seconds have elapsed between a flash of lightning and the next clap of thunder, the storm will be approximately four nautical miles away from your position, as 22 divided by 5.5 equals 4.

If you wish to determine the distance in kilometers or statute miles, you may assume that sound travels one kilometer in 3 seconds or one statute mile in 5 seconds.

RAIN

Sad to say, the sun does not always shine on recreational boaters as they romp over the bounding waves. They can expect to be bathed frequently with nature's gift from the heavens, rain. Essential as this gift may be to maintain the balance of nature on earth, most of the time it can turn a boating excursion into a downright dismal experience.

Nevertheless, rain is an important participant in the weather story. Air masses absorb moisture as they move over large bodies of water, so that a warm air mass crossing the Gulf of Mexico will pick up moisture, as will a cold air mass traveling over the Great Lakes. When the saturation point is reached within an air mass and the air temperature and the dew point temperature are nearly the same, condensation takes place, and this moisture is returned to the earth most frequently in the form of rain. Rain provides visible evidence of many different weather conditions such as the prolonged drizzle preceding a warm front or the heavy rain during a cold front.

As you have already learned, the possibility of rain can be predicted when certain cloud formations are observed. Other visual signs indicate the actual occurrence of rain. For example, when you can see a broad band of darkened sky descending to the earth's surface from a cumulonimbus cloud, you can be fairly sure that persons and objects exposed to the elements in that area are getting very wet. The approach of a line squall over the water is another visible and more alarming sign of rain for boatmen. The advance of this type of rain storm across the water appears more relentless and more vivid than it does on land. It looks like a vertical curtain of water coming toward you, and soon you are contending with the pelting, blinding rain, strong winds, and confused seas it brings with it.

FOG

Fog is condensed water vapor in the air that lies at or near the surface of the land or water. Fog can often roll in suddenly, obscuring familiar objects ashore before you are aware of it, especially when you are engrossed in fishing or conversation. From a distance, a fog bank rolling over the sea looks like a thick gray filling between a layer of sky above and water beneath. Looking toward land, the first sign of fog may be merely a blurring of the contours of trees and buildings.

Be warned that fog can play unbelievable tricks with your visual perception. One weekend, the Ericsons and the Townsends waited in Marblehead Harbor in Massachusetts for a morning fog to lift before setting off on a cruise to Cohasset, approximately a four-hour trip in the Townsends' auxiliary.

As the weather was questionable, we followed a longer course than usual in order to take advantage of available sound or audible buoys. We recorded the time on the chart whenever we could visually determine our position from land or sea marks. The adoption of this procedure proved to be wise because a dense fog closed in as we were crossing the mouth of Boston Harbor.

Virginia and Sallie were stationed on the foredeck when the entrance buoy to Cohasset Harbor was expected to come into view. When we heard its bell, we pointed toward the sound and the skipper headed in that direction.

Suddenly the bell popped into sight out of the woolly blankness. Our hearts leaped into our mouths because the bell appeared to be so tall that for a harrowing instant we all thought it was rockbound Minot's Ledge Light, a high tower that should have been well to port of us. We heaved a collective sigh of relief when we realized that we were seeing not only a hazy, artificially enlarged outline of the bell, but also its reflection in the mirrorlike surface of the water.

The water and air were so still and the light so strangely diffused that day that all previously familiar objects within our small area of visibility were magnified out of all proportion. Breadbox-sized lobster-pot markers loomed like large navigational buoys and the sea gulls were even more startling with their fairly modest proportions distorted into those of prehistoric birds pictured in encyclopedias.

In foggy weather, many persons who do their boating in places like the island-studded Maine waters make it a practice, whenever possible, to chart courses that take them near the wind-sheltered sides of islands. As the wind blows across an island, the warm air rising from the land tends to evaporate the fog temporarily and less fog or no fog at all is likely to be close to the shore of the side of the island that is sheltered from the wind.

The possibility that fog will creep in to smother your location with its damp, opaque blanket is determined by various combinations of the amount of moisture in the air, the temperature of the air, and the presence, direction, and velocity of the wind. This silent visitor has three personalities, which are identified as radiation fog, advection fog, and frontal fog. The names are clues to the specific combination of conditions necessary before any one type can materialize on the weather scene.

Radiation Fog Radiation fog, or ground fog as it is sometimes called, is of interest to boatmen because it drifts into harbors at night from the adjacent land. This fog forms only on clear nights and is especially likely to form in marshy lowlands.

Radiation fog usually forms during the evening when the land has cooled because the heat of the sun is no longer being absorbed by it. The radiating effect of the cool ground now cools the air just above it. When this newly cooled air is moist and its temperature reaches the dew point, condensation will occur. Only dew will be formed at this time if there is no wind. If, however, there is a wind of two or three knots, it will agitate this layer of surface air slightly so as to thicken it and cool it a little more than is necessary for the formation of dew alone, and radiation fog will be formed as well.

This type of fog can be expected to be "burned off" or evaporated early the next day by the sun, but the earlier it forms during the previous evening, the longer it will last after the sun has risen.

Radiation fog will not form when stratus clouds are present overhead in the evening because this thick cloud layer blankets the land allowing it to retain the heat it absorbed during the day.

Advection Fog Nearly all fog encountered at sea is advection fog. Advection means the horizontal movement of heat from one place to another, and this fog is formed at sea when warm air is blown over cold water by winds with speeds of 4 to 15 knots.

This type of fog is difficult to forecast as it may be formed many miles away and blown toward a coast by the wind. It may persist for several weeks and be either dense or patchy, but it is not likely to dissipate completely until there is a change in the direction of the wind.

Advection fog commonly occurs on the west coast of the United States when warm air is blown over the cold California Current and on the northeast coast when moist tropical air over the Gulf Stream encounters cold air over the Labrador Current.

Frontal Fog Frontal fog is often produced during the approach of a warm front when moist warm air rises over a cooler air mass. This occurs because the moisture in the air increases when warm air mixes with cooler air. This type of fog will generally remain in an area until the warm front passes.

WIND CATEGORIES

Wind may seem to be an unlikely topic to include under visual signs, but it belongs here because you can see its effect on things around you as well as feel it on your body. However, the force of air that whips spray into your eyes, snaps a flag to attention, or propels a sailboat through the waves is only part of the overall wind story. We define the role of wind in the mystery story that weather often seems to be by discussing three types of winds. These are the great winds called terrestrial winds, the familiar winds referred to as surface winds, and the sea and land breezes that we call local winds.

Wind Directions Basic to any discussion of wind is the fact that all winds are identified by the direction *from* which they are blowing. The cardinal points or principal wind directions with their equivalent compass degrees are as follows:

Cardinal Points	*Degrees*
North (N)	000° or 360°
East (E)	090°
South (S)	180°
West (W)	270°

These directions can be further divided with northeast (NE) as 045° and southeast (SE) as 135° and so on. The old-time sailors would break this down further into 32 points of the compass with one point equaling 11¼ degrees, but degrees are used almost entirely now for more precise wind directions.

Terrestrial Winds Terrestrial winds are the great upper-level winds of the earth. They carry along with them the weather systems we experience on earth, and, for this reason, the paths of some of these winds are important to our study of weather system movement in different sections of the United States.

Basic to the understanding of the paths of terrestrial winds is the premise that warm air rises from the earth and cool air sinks to it. Thus, these great winds are a product of the extreme difference in air temperature between the hot equator and the cold poles. Theoretically, terrestrial winds are caused in part by the motion of the hot equatorial air rising into the atmosphere as a result of the continual heating action of the sun's rays on the torrid zone around the middle of the earth. This air divides high above the earth to flow toward the frigid poles, half toward the North Pole and half toward the South Pole. As the air reaches the regions of the poles, it is chilled and descends toward the surface of the earth to flow back to the equator where the cycle is begun anew. This is an oversimplification, of course, but it serves our purpose for this discussion.

Since the spinning motion of the earth during its 24-hour daily rotation is greatest at the equator and nonexistent at the poles, the terrestrial winds are deflected to follow different paths during the round-trip excursions. There are two paths that affect the movement of weather systems over the United States, and the terrestrial winds that follow these are called the Prevailing Westerlies and the Northeast Trade Wind (although this trade wind is commonly referred to as the Northeast Trades, it is also called the Easterly Trades along the east coast of Florida and the Gulf of Mexico because the prevailing winds blow from the easterly quadrant). The Prevailing Westerlies flow generally from west to east across the continental United States north of latitude 30°N. South of this latitude, the Easterly Trades flow generally from east to west in the region that affects the United States.

Since latitude 30°N is considered to be the approximate dividing line between these two great wind systems, the portion of it pertinent to our study stretches from the Atlantic Ocean coast of northern Florida, west across the upper rim of the Gulf of Mexico, through the Republic of Mexico about 150 nautical miles south of California, and then on across the Pacific Ocean to the north of Hawaii.

In the days when all ocean commerce was carried on by vessels propelled solely by sail, mariners discovered that ocean crossings were far less time consuming when the westbound voyages were made in regions influenced by the Easterly Trades and eastbound voyages, in regions influenced by the Prevailing Westerlies.

Surface Winds Developing within the terrestrial wind systems are a continuous series of high and low air pressure areas that form near the earth's surface and are carried across the continental United States by the Prevailing Westerlies. The air spiraling into these pressure areas is the surface wind we feel on land and sea. In the Northern Hemisphere, air spirals out from the center of a high pressure area in a clockwise direction to flow in toward the center of a low pressure area in a counterclockwise direction.

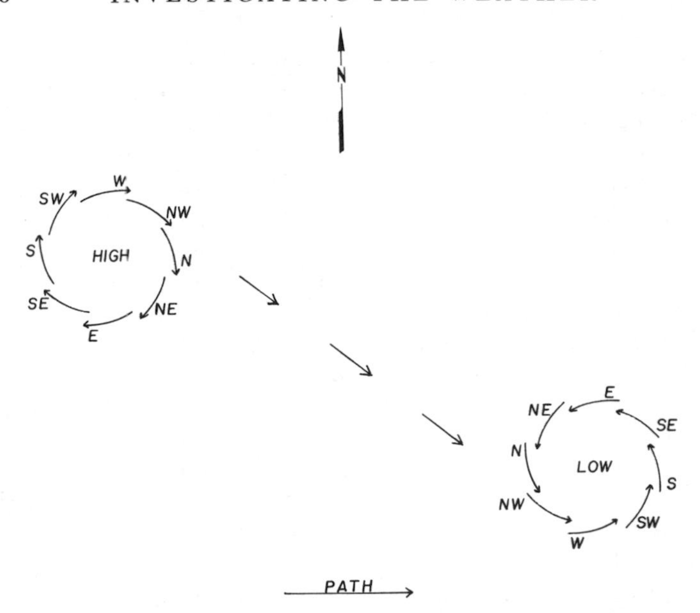

Fig. 4. WIND FLOW IN PRESSURE AREAS—Northern Hemisphere *Wind indicator arrows are labeled with the direction from which the wind is blowing: wind spirals clockwise out from the center of a high and counterclockwise in toward the center of a low.*

Surface wind direction is indicated by the wind arrow that points toward your location from the pressure system dominating your region's weather.

Buys-Ballot's Law points out that if you face into the wind, the center of a low pressure area is to your right and slightly behind you.

Meteorologists refer to the movement of air out from the center of any high as anticyclonic circulation and the movement of air toward the center of any low as cyclonic circulation. When meteorologists identify a storm as a cyclone, the term is used solely to describe the pattern of wind circulation within a storm; it has nothing to do with the type or severity of a storm.

Surface winds between the centers of highs or lows increase in direct proportion to the rapidity of barometric air pressure change, as you have seen from your study of isobars on weather maps. The direction from which the surface wind will be blowing at your location is directly related to your position with respect to that of a high or low pressure area.

Buys-Ballot's Law The position of a low pressure area tive to your location for any one moment and whether it is com g toward you or has passed you by can be determined by applying Buys-Ballot's Law. This law is sometimes referred to as the "Right Hip Pocket Rule" because when you face into the wind in the Northern Hemisphere, the center of the low pressure area is to your right and slightly behind you. Nautically speaking, the low could be said to bear 90° to 120° from the direction in which you are facing.

Since weather systems generally move from west to east (in the middle latitudes) across the United States, the following propositions of the law should be valid for this area of the earth's surface.

1. Where there is a north wind (360°), the center of the low has passed over you. (To demonstrate this, assume you are standing on the north wind line depicted to the left of the low on Fig. 4, page 30, and face into the wind.)
2. When there is an east wind (090°), the center of the low is going by to the south of you.
3. When there is a south wind (180°), the center of the low is coming toward you.
4. When there is a west wind (270°), the center of the low is going by to the north of you.

Surface Wind Shifts When the direction from which the wind is blowing changes, the wind is said to either back or veer. When the wind backs, the direction from which it blows shifts counterclockwise or backward through the degrees of the compass. Alternatively, when the wind veers, it shifts clockwise through the degrees of the compass.

Because of the counterclockwise flow of air in toward the center of a low pressure area in this hemisphere, the surface wind will shift in your local area as the center of a low passes over you or passes by either to the north or south of you. You can expect one of the following wind shifts to occur when you know the position of the center of a low relative to your location.

1. When the center of the low passes to the south, the wind will back from approximately east to a northerly or northwesterly direction or from 090° to 360° or 315°.
2. When the center of the low passes over you, the wind will change from approximately southeast or south to a northwesterly or northerly direction or from 135° or 180° to 315° or 360°.

3. When the center of the low passes to the north, the wind will veer from approximately southwest to a northwesterly or northerly direction or from 225° to 315° or 360°.

Local Winds: Sea Breeze and Land Breeze Sea and land breezes are the names for gentle to moderate local winds occurring along a shore. They originate somewhat in the same way as the terrestrial winds, although the process is on a very much smaller scale and is not as continuous.

As the land near a fairly sizable body of water heats up on a hot day, the warmed air will rise and the cooler air over the water will flow toward the land to replace it. This phenomenon is called an onshore breeze or a sea breeze.

During the night following a hot day, the land cools more quickly than the water and the reverse situation occurs. As the warmer sea air rises over the water, the cooler land air blows toward the water to replace it creating a land breeze or offshore breeze.

If the day is hot and there is little wind over the water for passage along a seacoast, a sailor can anticipate the arrival of a sea breeze sometimes as early as noon to speed him along with a pleasant beam wind. The effect of this is usually limited to less than five nautical miles from shore.

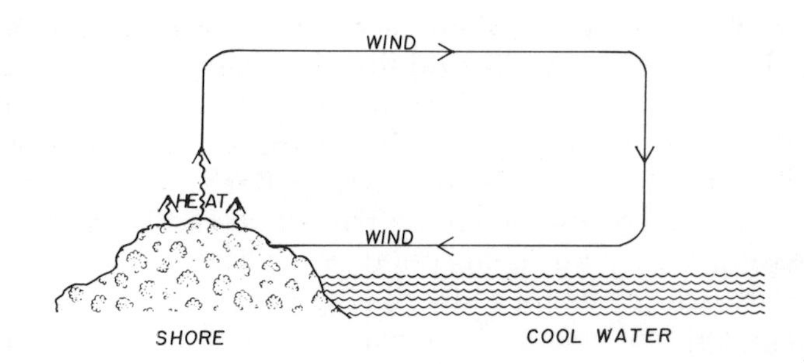

Fig. 5. SEA BREEZE *Development of an on-shore breeze during a hot afternoon.*

Wind Signs It is difficult to determine the direction of the surface wind when you are aboard a moving boat because the wind you feel is a combination of wind resulting from the boat's movement through the air and the surface wind. You can still estimate surface wind direction and sometimes even its speed by making the following visual observations:

Surface of the water. The wind will be nearly perpendicular to the line of the waves. If there is enough wind to blow spray off the top of the waves, the direction of the surface wind is indicated more clearly. It must be noted, however, that swells, as opposed to waves, can be caused by leftover seas from a past or distant storm and do not have any relation to the direction of the wind.

Low clouds. The direction of the movement of the lowest layer of clouds is an indication of the direction of surface winds. Do not be fooled by the movement of high clouds as they might be in an entirely different wind system.

Smoke on shore. The direction of blowing smoke is a clear indication of the direction of surface winds. Many a sailboat race has been won by an alert skipper who noticed an approaching wind shift by the change in direction of distant smoke.

Flags. Flags on shore or aboard anchored boats are often a good indication of the strength of the surface wind except when the wind directly affecting these flags has been deflected by surrounding terrain or buildings. A flag starts to lift when the wind is blowing about five knots. According to our observations, when compared with National Weather Service reports concerning local wind speeds, flags are likely to be extended when the wind is blowing 8 to 12 knots. When the wind is blowing 25 knots or more, flags are fully extended with a pronounced upward lift.

Beaufort (Boh'-fort) Scale of Winds The Beaufort scale is a series of numbers set up in 1806 by Sir Francis Beaufort, a British admiral, to define increasing strengths of wind. The classifications in his scale range from force 0 for calm air to force 12 for the onset of hurricane winds. The use of Beaufort's number code by telegraphers to transmit wind conditions at sea became widespread by the middle of the nineteenth century. Descriptive data pertinent to the relationship between wind force, the state of the offshore sea, and between the wind force and various signs on land have been added from time to time to the original definitions. In Fig. 6, page 34, are excerpts from a table in the *American Practical Navigator** by Nathaniel Bowditch listing the Beaufort scale force numbers with their corresponding knots and kilometers per hour of wind speed and visual signs for estimating the wind speeds for the effects observed at sea and on land.

* Published by the Defense Mapping Agency, Hydrographic Center (DMAHC), Washington, D.C. 20309; may be ordered directly or purchased from a nautical supply store.

BEAUFORT SCALE

Force No.	Knots	Estimating Wind Speed	
0	under 1	Sea:	Sea like mirror.
		Land:	Calm; smoke rises vertically.
1	1-3	Sea:	Ripples with appearance of scales; no foam crests.
		Land:	Smoke drift indicates wind direction; vanes do not move.
2	4-6	Sea:	Small wavelets; crests of glassy appearance, not breaking.
		Land:	Wind felt on face; leaves rustle; vanes begin to move.
3	7-10	Sea:	Large wavelets; crests begin to break; scattered whitecaps.
		Land:	Leaves, small twigs in constant motion; light flags extended.
4	11-16	Sea:	Small waves, becoming longer; numerous whitecaps.
		Land:	Dust, leaves, and loose paper raised up; small branches move.
5	17-21	Sea:	Moderate waves, taking longer form; many whitecaps; some spray.
		Land:	Small trees in leaf begin to sway.
6	22-27	Sea:	Larger waves forming; whitecaps everywhere; more spray.
		Land:	Larger branches of trees in motion; whistling heard in wires.
7	28-33	Sea:	Sea heaps up; white foam from breaking waves begins to be blown in streaks.
		Land:	Whole trees in motion; resistance felt in walking against wind.
8	34-40	Sea:	Moderately high waves of greater length; edges of crests begin to break into spindrift; foam is blown in well-marked streaks.
		Land:	Twigs and small branches broken off trees; progress generally impeded.
9	41-47	Sea:	High waves; sea begins to roll; dense streaks of foam; spray may reduce visibility.
		Land:	Slight structural damage occurs; slate blown from roofs.
10	48-55	Sea:	Very high waves with overhanging crests; sea takes white appearance as foam is blown in very dense streaks; rolling is heavy and visibility reduced.
		Land:	Seldom experienced on land; trees broken or uprooted; considerable structural damage occurs.
11	56-63	Sea:	Exceptionally high waves; sea covered with white foam patches; visibility still more reduced.
		Land:	Very rarely experienced on land; usually accompanied by widespread damage.
12	64+	Sea:	Air filled with foam; sea completely white with driving spray; visibility greatly reduced.
		Land:	Very rarely experienced on land; usually accompanied by widespread damage.

Fig. 6. BEAUFORT SCALE *Following are the kilometers per hour–knots figures from the U.S. Hydrographic Agency Beaufort scale. Published by the U.S. Hydrographic Agency.*

Force No.	*km/h*	*knots*
0	under 1	under 1
1	1–5	1–3
2	6–11	4–6
3	12–19	7–10
4	20–28	11–16
5	29–38	17–21
6	39–49	22–27
7	50–61	28–33
8	62–74	34–40
9	75–88	41–47
10	89–102	48–55
11	103–117	56–63
12	118+	64+

Wind Warnings: Flags and Lights The Beaufort scale per se is not used by the National Weather Service but their wind warnings are based on the fresh to strong winds of the scale forces 5 to 7, gale winds of forces 8 and 9, storm winds of forces 10 and over, and hurricane winds of force 12 or higher. For years, flags and lights were posted along the coastline to warn boatmen of strong wind conditions. This was an outgrowth of a railroading custom in effect over a hundred years ago when flags were flown from trains to warn residents along the tracks of imminent storms.

The National Weather Service informs us that the posted wind warning system for boats is being phased out. They feel this type of weather warning may be misleading and even dangerous because they cannot guarantee that the signals will be displayed in time or discontinued when they are no longer valid. The Weather Service prefers instead to train the boating public to utilize the VHF-FM Continuous Weather Broadcasts (see page 68).

Because visual signals are still being displayed at some locations, however, we feel we should include the following descriptions of these warnings along with their expected wind speeds:

Small craft advisory: one red pennant flown during the day and a red light displayed over a white light at night. Winds 18 to 33 knots.

Gale warning: two red pennants flown during the day and a white light displayed over a red light at night. Winds 34 to 47 knots.

Storm warning: one square red flag with a black square in the center flown during the day and two red lights displayed one over the other at night. Winds over 48 knots.

Hurricane warning: two square red flags with black squares in the center flown one above the other during the day and two red lights, one above and one below a single white light displayed at night. Winds of 64 knots and above in association with a hurricane.

SEA CONDITIONS

Although wind is only one of the forces that can be responsible for the formation of sea conditions, its influence is often the most detectable. Among the other forces contributing to sea conditions are tidal currents and ocean currents as well as infrequently occurring phenomena like those discussed in Chapter 11 under tsunamis.

Following are some of the terms used to describe the surface form of a wave and a few simplified explanations of the more familiar sea conditions. These are intended to serve only as a brief introduction to this complicated subject.

Wave progression: unseen energy from the wind is transferred to the water and advances through it in much the same way a disturbance travels through a long rope when one end is flicked up and down. The movement undulates through the rope, but the rope itself does not move forward. Wave formations on the surface of the water are the visible evidence of this energy.

Crest: the top part or ridge of an individual wave.

Trough: the lowest part or bottom between two successive crests.

Wavelength: the horizontal distance between two successive crests.

Height: the vertical distance between the crest and trough of an individual wave.

Steepness: the ratio of wave height to wavelength.

Fetch: the distance the wind blows in one unchanging direction. The highest waves build up when there is a strong wind of long duration blowing over a long fetch.

Deep-water waves: waves that advance through water that is deeper than one-half their wavelength. Water particles near the surface of the water move in one complete elliptical orbit with the passage of each wind wave to return nearly to their original positions. The circular disturbance decreases in width with depth and is considered to be negligible in water deeper than one-half the wavelength of the wave.

Shallow water waves: waves that advance over water less than one-half their wavelength. Here friction with the bottom changes the orbital wave form.

Swells: waves that move out from the high winds that cause them. As swells outrun the storm, their profile rounds out; their wavelength increases and their height decreases. Swells can travel for thousands of miles.

Height increases in waves: this basically depends on the strength of the wind, the length of time the wind has been blowing, and the distance of the fetch. Following is an extension of the Beaufort scale that includes the wave terms used to describe the heights likely to occur in a fully developed sea at various wind speeds. (The knots, kilometers per hour, and feet measurements are the figures given in the U.S. Hydrographic Agency tables; we have added the conversions to meters.)

Force	*Wind Speed in*		*Sea Terms*	*Height of Waves*	
	Knots	*Kilometers per hour*		*Meters*	*Feet*
0	under 1	under 1	Calm	0	0
1	1–3	1–5	Smooth	less than 0.3	less than 1
2	4–6	6–11	Slight	0.3–0.9	1–3
3	7–10	12–19	Moderate	0.9–1.5	3–5
4	11–16	20–28	Rough	1.5–2.4	5–8
5	17–21	29–38	Rough	1.5–2.4	5–8
6	22–27	39–49	Rough	1.5–2.4	5–8
7	28–33	50–61	Very rough	2.4–3.7	8–12
8	34–40	62–74	Very rough	2.4–3.7	8–12
9	41–47	75–88	High	3.7–6.1	12–20
10	48–55	89–102	Very high	6.1–12.2	20–40
11	56–63	103–117	Mountainous	12.2 and higher	40 and higher
12	64+	118+	Confused		

Mariners tend to overestimate wave heights. Few waves attain more than 17 meters (55 feet) in height and before this can happen the wind has to be blowing for 24 hours at about 60 knots over a 450-nautical-mile fetch. The heights attained are limited by the tendency of waves to crest and break at various critical heights. Winds of hurricane strength blow off the tops of waves and the height decreases when waves outrun a storm and become swells. However, in highly confused seas, individual peaks can reach much greater heights than other waves in the vicinity. There is well-documented evidence of the occurrence of single waves that have attained heights of 30 meters (100 feet). Presently unpredictable, these so-called super or killer waves are the subject of intensive international studies.

Confused seas: seas where waves or sea conditions are no longer orderly. Confused seas build up from the interaction of different wave systems. For example, this can occur when local wind waves meet ocean swells at an angle. They are also likely to develop in relatively narrow passages or when obstructed by land.

Strong tidal currents can cause this type of disturbance with steep irregular seas developing when the wind blows against the flow. Agitated water, sometimes accompanied by swirling eddies, can develop when the path of a current is deflected by a land mass such as an island or a jut in the coastline. Confused wave formations called tide rips result either from the convergence of opposing currents or from the action of a strong current flowing over a shoal, irregular bottom surface. The friction with the bottom in shoaling waters tends to refract the waves and cause them to parallel the shore, to increase in height, and to break.

Breaking waves: waves whose crests topple over into curls of frothy water. In theory, deep-water waves break when their height is greater than 1/7 their wavelength; however, they have often been observed to break at 1/10 to 1/12 their wavelength. Notice in the Beaufort scale that the first whitecaps form at sea with winds of only 7 to 10 knots.

In shallow-water waves, the friction of the bottom causes the wave formations to be compressed. As the wavelengths decrease, the heights increase, resulting in steep waves and breakers that warn either of shallow water over rocks and sand bars or of the more gradual shoaling that occurs near a coast.

When ocean swells reach shallow water, their wavelength decreases and even formerly gentle rollers will steepen to break over the shoals.

Surf: a succession of breaking waves along a shore.

Bar conditions: breaking or confused seas resulting when an opposing wind or ocean swells meet an outgoing current at the entrances to rivers and inlets or the entrances of harbors where sandbars have built up. Normally, there is a period of about two hours once or twice every 24 hours when this condition can be expected. When there are swells, tidal current conditions, even on the calmest days, may be such that small boats can be endangered.

In some, but not all, areas where bar conditions are severe, bar-guide advisory signs have been established on which amber lights flash alternately when the bar conditions are considered dangerous. However, there is no guarantee that a bar is safe when the lights are not flashing.

Some additional information on sea conditions is contained in the following excerpts from "Some Interesting Points on Waves and Swells," a paper written by Robert E. Lynde, marine forecasting specialist, National Weather Service, Boston, Mass.:

> A mariner's rule of thumb that relates wave height to wind velocity says that the height ordinarily will not be greater than ½ the wind speed—thus an 80 mph wind could produce 40-foot waves.*
>
> At a water depth of ⅓ the wave's height, the wave will become unstable and break. The result is breakers or surf.
>
> Wind drifts (currents produced by the force of the wind), ocean currents and tidal currents affect the waves. Briefly, a contrary current will decrease the wave length and increase the height of the waves, making them steeper. Following currents have the opposite effect: they increase the wave length, decrease the wave height and flatten the sea. In general, the stronger the current the greater are its effects upon the waves.
>
> On the shelf waters, tidal currents are strong enough to steepen waves. Tides run with velocities as high as 1.3 knots around Nantucket Shoals (Massachusetts) and up to 4 or 5 knots in restricted coastal areas such as between some islands in Maine. These conditions can transform a very moderate sea into a very dangerous one for small craft when waves are running against the tidal current. Names such as Pollock Rip, Race Point, etc., are example of local areas that derive their names from these conditions.

---

* 80 mph = 70 knots; 40 ft. = 12 meters.

CHAPTER 3

Detect Weather with Instruments

Meteorologists use many instruments in their all-out effort to track down and study the elusive and often mysterious ways of the atmosphere. The instruments we have chosen for this chapter, however, have been selected as those most likely to be found in the homes of amateur weather enthusiasts whose interest stems mainly from their association with boating and their desire to check on the weather before starting out on the water. Persons with cruising boats often have duplicates of many of these instruments aboard their boats as well as at home in order to keep abreast of weather changes.

BAROMETER

A barometer measures atmospheric air pressure and is a helpful tool for predicting weather conditions. A single reading on a barometer is not always informative by itself, however. The trend of barometric measurements made within a three-hour time period provides the best clue to weather changes.

The passage of a front is detectable from barometric readings. Because fronts are troughs of lowest barometric pressure in an area, your barometer will drop and then rise as they pass through your location. If the atmospheric pressure, as shown on a barometer, falls rapidly in the middle latitudes a strong cold front is not far away, whereas an intense storm in the southern sections of the country will be indicated by the instrument with only a slight change. The barometer rises with the approach of a high pressure area, and the amount of change is also influenced according to the latitude in which it is located. You can expect present weather conditions to prevail for a while longer when your barometer is steady.

Frontal passages are more pronounced in the northern portions of the United States. Pressure differences between highs and lows gradually moderate the farther south you go; changes indicated by a barometer are much less obvious in areas such as Florida and the gulf coast. Here you must watch the instrument very carefully in order to observe any change in overall trend.

Each day barometers also register relatively small tidelike variations of air pressure, the cause of which is not completely understood. These are referred to as diurnal changes and are not associated with either high or low pressure areas. Diurnal changes are usually less than 0.34 kiloPascals (3.39 millibars or 0.10 inches of mercury). They have peaks about 10 A.M. and 10 P.M. and dips at 4 A.M. and 4 P.M. local standard time. These normal fluctuations are more pronounced in the tropics than in the middle and higher latitudes.

This phenomenon is graphically illustrated on tracings of barometric measurements that are made by a recording barometer called a barograph. In Fig. 7 (see page 44), two of our barograph charts are reproduced to illustrate not only the diurnal fluctuations of two widely-spaced latitudes, but also to compare how the atmospheric air pressure differs within storms of similar strengths in these different latitudes.

A mercurial barometer is a type of barometer that measures atmospheric pressure by means of mercury sealed into a glass tube marked with a scale. When a high pressure area is approaching, the pressure in the atmosphere increases, forcing the mercury to rise in the tube. This accounts for the legendary seafaring expression "The glass is rising," which still has the same happy weather connotation it did in the days of yore when shouted by a jubilant captain to encourage his crew, because he knew it indicated his sailing ship had successfully survived a treacherous storm at sea. Conversely, when a low pressure area is en route, atmospheric pressure decreases and the mercury falls, thus showing a lower reading on the scale.

Although it is not quite as accurate as the mercurial barometer, the mechanical aneroid barometer is the type found most often on cruising boats of today because it is sturdier and more compact. An aneroid barometer has an expansible chamber containing a partial vacuum and including a flexible diaphragm or bellows. When exposed to the atmosphere, the chamber expands and contracts as the atmospheric pressure falls or rises respectively. The chamber is connected to one end of a needle through an intermediate linkage system. As the chamber is compressed or expanded by changes in the air pressure of the atmosphere, the needle will swing to point to the appropriate measurement figure marked on the face of a circular dial.

An aneroid barometer should be tapped prior to reading because of the tendency of mechanical parts to stick. An additional adjustable needle provided on most of these instruments can be turned by hand at the time of the reading so that it is positioned over the registering needle. This marks the last reading of barometric air pressure so that any change can be noted quickly the next time you look at the barometer. Aboard a cruising boat, the barometer reading is often recorded at regular intervals in the ship's logbook.

Your barometer should be corrected to give barometric air pressure readings at sea level for the place where your instrument is located. Once this is done, it is possible to make a realistic comparison of your barometric air pressure measurement with that given on a weather broadcast for areas westward of your location in order to predict the weather that is probably coming in your direction.

In addition, when your barometer is corrected, you will be able to tell whether you are in a high or low pressure system according to the standard atmosphere figure on your instrument. Standard atmosphere figures as well as the atmospheric pressure conversions for kiloPascals, millibars, and inches of mercury are given in the Appendix.

Barometric air pressure decreases as height above sea level increases. The rate of this decrease is approximately one-hundredth of an inch of mercury (0.01 in.) for every 10 feet of elevation.* Although this figure may seem small, measurements of barometric pressure in units of inches of mercury are given in station reports for every hundredth of an inch. As an example of the magnitude of this change, if your home is only 305 meters (1,000 feet) above sea level, there is a decrease in the atmospheric pressure of one inch. This is significant when you consider that many U.S. customary barometers are scaled from 28.5 to 31 inches of mercury, only a 2½-inch span.

You will be able to adjust your barometer to sea level by listening to weather reports, but only if you can be sure that the barometric reading given was reported from a nearby weather station and, according to the announced time of the reading, was made within the past hour. This type of current barometric information is given sometimes by weathercasters on television programs. It is also obtainable at the

* The rate of decrease is 0.34 millibars for every 10 feet of elevation; this is the same as 0.03 kiloPascals for every 3 meters of elevation.

beginning of a taped National Weather Service VHF-FM Continuous Weather Service broadcast.

If you have a radio capable of receiving Aviation Weather Reports from a nearby aircraft reporting station, you can use the altimeter setting figure, which is customarily given hourly, for your barometric reading. The Federal Aviation Agency's publication *Private Pilots Handbook* explains this as follows:

> Since the rate of decrease in atmospheric pressure is fairly constant in the lower layers of the atmosphere, the approximate altitude can be determined by finding the difference between pressure at sea level and pressure at the given altitude. In fact, this is the principle upon which the airplane altimeter operates. The scale on the altimeter, instead of indicating pressure in terms of inches of mercury, indicates directly in terms of feet of altitude.

The pilot sets his altimeter for the latest barometric reading at his destination airport.

BAROGRAPH

A barograph is an aneroid barometer that provides a visual record of barometric air pressure recorded with a continuous line on a chart of graph paper by means of a sensitive pen arm. The chart, which is labeled across the top of the graph with the days of the week, is wrapped around a drum rotated slowly by a clock mechanism. While a barograph is more expensive than some other types of barometers, we believe it is well worth the difference in order to have a record of air pressure trends.

There are three types of barographs: those that require an electric outlet, those operated by batteries, or those that must be wound by hand once a week. Even the hand-wound type is just as convenient and easy to wind on a regular basis as it is also necessary to change the graph paper weekly. Refills of graph paper are sold in yearly supplies of 52 sheets to a packet and can be ordered either directly from the manufacturer or from a nautical supply store.

During rough weather when the boat slams into heavy seas, the pen arm of the instrument is likely to jiggle wildly up and down. To help alleviate the problem, Sam Townsend, Jr., designed and constructed a spring mounting for the Townsends' barograph. We include his design should you wish to construct a similar mounting.

He cut a rectangular wooden platform out of 5/16-inch teak to the measurements of the bottom of their barograph. The cover of the instrument's case was opened in order to drill holes in the bottom of the

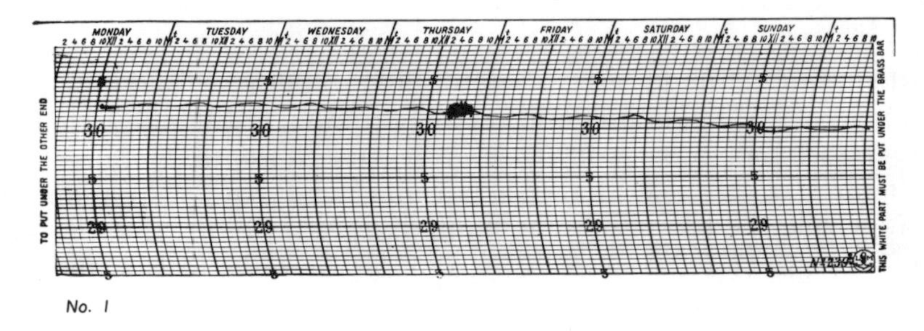

No. 1

No. 2

Fig. 7. BAROGRAPH CHARTS *Sample graph papers of storms of about equal intensity as recorded in different locales.*

No. 1 Recorded in the Bahama Islands at about the same latitude as Miami, Florida, during January.

Thursday afternoon was a particularly rough passage. It was blowing about 25 knots and the seas were 1.8 meters (6 feet) or higher.

The boat was in port during a severe storm Sunday afternoon and evening when winds were over 50 knots and gusting. Note that this storm is hardly noticeable on the barograph, whereas the 10:00 A.M. *and* P.M. *daily peaks and 4:00* A.M. *and* P.M. *daily dips are very pronounced in this tropical area.*

No. 2 Recorded inside in Marblehead, Massachusetts, during January. Monday and Saturday storms both had gusting winds of over 50 knots. The sharp slopes toward the centers of the lows are very pronounced on this chart, while it is difficult to distinguish the 10:00 A.M. *and* P.M. *peaks and the 4:00* A.M. *and* P.M. *dips.*

case through which the barograph could be screwed to the platform. The screws do not show when the top of the case is closed.

The teak platform is freestanding, supported at each corner on a shelf by a ¾-inch noncorrosive coil spring about 1-inch-high. Each spring has a cube of foam rubber jammed into it to damp its action. The shelf is constructed of 5/16-inch-thick teak and has a 1½-inch-high wooden railing around the front and sides to hide the coils. The platform to which the barograph is attached is set in ½-inch on all sides

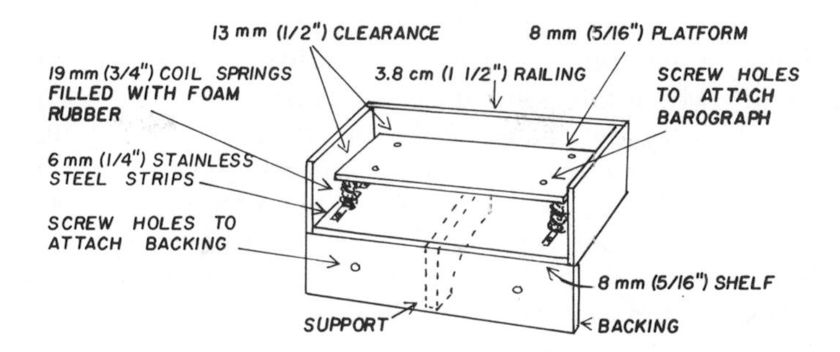

Fig. 8. BAROGRAPH MOUNTING *A spring mounting to absorb shocks aboard a boat to lessen vibration jiggle in the pen arm.*

According to an official directive concerning lumber thickness, the industry does not have to retool to metric measurements. Instead, the nearest whole number of metric measurement is used to describe the customary thickness of plywood or lumber.

from the railing and the back of the shelf so that it can move freely on its coils.

The ¾-inch coil springs are attached to the bottom of the platform and to the top of the shelf through 4 shallow, dead-ended holes, ¾-inch in diameter and 1/16-inch deep, drilled in both the platform and the shelf and into which the ends of the springs fit snugly. The end coils of the springs are held in place in these shallow holes with thin strips of stainless steel ¼-inch wide and 1¾-inches long. Holes were drilled through each end of these strips so that they could be screwed to the freestanding platform and the shelf.

The underside of the shelf is attached with a knee support to a backboard, which in turn is screwed to a bulkhead of the boat. An advantage of this mounting is that the barograph can be removed easily from the platform at any time, such as when a boat is laid up for the winter.

ANEMOMETER

An anemometer is an instrument that measures the speed of the wind and can be very helpful in deciding whether it is sensible to go out on the water. Furthermore, a comparison of wind speeds reported from weather stations to the west of your location with local wind speeds can alert you to approaching weather conditions.

The small rotating wind cups or concave blades found on most of these instruments must be installed at the highest point of a boat or house so that the wind spins them freely with as little obstruction to free wind flow as possible. Wind speed is indicated on a dial according to the speed at which the wind spins the cups.

A less sophisticated hand-held wind indicator is available, although obviously it is often difficult to hold it high enough to ensure free wind. At first glance, the pocket-sized anemometer we are familiar with resembles a common household thermometer, but instead of having fluid in a sealed tube there is a small, freely moving ball and two small openings at the base of the back of its tube. This instrument is held in an upright position with the two openings toward the wind, and the action of the wind blowing through these openings forces the little ball up the tube. The speed of the wind is indicated by the ball's position in relation to the wind speed measurement scale on the side of the tube.

Aboard a moving boat, the apparent wind speed, which is a combination of the surface wind and the wind caused by the boat's movement, is indicated on the anemometer dial. On a flat, calm day, the wind speed registered will be equal to the speed of the boat. It is necessary, when it is windy, to apply the speed of the boat to the wind speed reading in order to determine just how strong the wind is blowing. For instance, when the wind is astern, you have to add your boat speed to your anemometer reading to find out the actual speed of the surface wind. If your boat is moving to windward or into the wind, you will have to subtract your boat speed to determine a realistic figure for the wind's speed.

An anemometer is very helpful aboard a sailing boat as it is possible to decide from the wind speed registered the best combination of sails to use.

Surprisingly enough, having an anemometer aboard your boat can be a social plus. The Townsends and the Ericsons discovered the great appeal of this instrument in the Florida Keys when they were safely ensconced in a marina during a strong blow. It wasn't long before boating neighbors began to drop in to read the *Troubadour*'s anemometer dial and exclaim over the strength of the wind. Wind speed broke the ice and a very pleasant party resulted.

WIND DIRECTION INDICATOR

Wind direction indicators rotate with the wind and point into it. They should be located as high as possible in a place where the wind flow is unobstructed.

Wind direction is given in true degrees in weather reports and forecasts. True north is located on surface maps at the North Pole or latitude 90°N where the longitude lines converge.

When you hear a weather report, you can tell from the direction the wind is blowing at a particular location whether or not a cold front has passed through that area. One of the most telling pieces of information provided by a wind vane is when it abruptly swings to indicate a northwest or north wind as a front passes through your area.

On land, the direction the wind is blowing can be determined by a weathervane on a roof. The arrow, rooster, or ship of the vane rotates freely on a support fixed with reference to the points of the compass; north should be positioned to indicate true north. The fixed support usually consists of just the cardinal points: N for north, E for east, S for south, and W for west. If the vane has a remote indicator wired to register inside the house, the dial will be marked either with the points of the compass or with a 360° scale.

It is not easy to obtain true wind directions on a floating boat. There you are oriented to compass directions by readings from the ship's compass. As most recreational boats have magnetic compasses, readings must be revised to true degrees in order to compare your wind direction observations with those given on weather broadcasts, even if your compass is correctly adjusted at all points in relation to magnetic north (magnetic north is situated approximately at latitude 74°N and longitude 101°W).

The easiest way to tell the direction the wind is blowing across the water is to head your boat into the eye of the wind by using a rotating wind indicator placed high on the boat as a guide. The needle on your compass will then point toward the wind and you will have its direction in magnetic degrees.

Surface wind direction is more difficult to determine when a boat is not heading directly into it or is dead before it. When this is the case, a boat's forward motion changes the direction of the surface wind experienced aboard. The combination of boat wind and surface wind is called apparent wind; the apparent wind direction is reckoned relative to the bow of the boat. The wind direction felt on a moving boat is always nearer to the bow than it would be when the boat is dead in the water.

Remote relative wind indicators can be purchased on which the readout is in the degrees of the relative apparent wind for either side of the boat as determined by the rotation of the wind indicator. The dial for the relative indicator is marked with zero degrees for the bow; each side is marked to 180 degrees. These relative readings are not to be confused with the magnetic degree readings on a ship's compass.

When the apparent wind is blowing toward your starboard side, the relative degrees registered on the wind indicator readout are added to the magnetic degrees of your compass; if the answer obtained is more than 360°, subtract 360. Relative degrees from a wind blowing over the port side are subtracted.

You are not completely out of the wind detection business if you do not have some type of mechanical instrument aboard. One of the best visual methods to determine wind direction is by observing the surface of the water since the crests of waves or wavelets can be considered to travel perpendicular to the wind. The wet-finger method may be even less precise but it is always available to feel the apparent wind direction. In both cases, use the ship's compass to find the approximate magnetic degrees for the direction from which the wind is blowing.

THERMOMETER

A thermometer, which measures the temperature of air or water, is the most universally known of all instruments used to gather weather information. However, your thermometer by itself only informs you of weather that has already arrived, such as the comparatively higher reading of a warm air mass and the lower temperature of a cold air mass; of course you can judge this less precisely for yourself by standing in the open air. As is the case with the other weather instruments discussed, a comparison of the trend between temperature readings in your location and reports of readings to the west of you is another aid in detecting the next weather act that is likely to appear in your area.

Because of its daily rotation, the earth cools at night when it no longer absorbs and radiates heat from the sun as it does during the day. Land is more sensitive to the cooling process at night and the warming process during the day than water, as demonstrated by the occurrence of sea breezes and land breezes. The temperature of the air decreases as it rises higher and higher into the atmosphere.

Most thermometers have mercury in a sealed glass tube with a small bulb at the bottom. As the air temperature increases, the mercury expands and rises along a scale marked in degrees on the outside of the tube.

Although for many years, in the United States and Canada, the majority of the scales on thermometers used for measuring surface temperatures have been marked in Fahrenheit (F) degrees, scales marked with the metric units of Celsius (C) degrees were used more widely internationally. In 1975, this international measurement was adopted in the United States, and during the transition period from Fahrenheit to Celsius, weathercasters often report air temperatures in both units. Conversion tables are given in the Appendix.

Some thermometers are especially designed to be immersed in water. Many fishermen rely on readings of a submersible thermometer not only to find out where to fish, but also how deep, because they know that certain types of fish are selective about water temperature. We find it desirable to keep a submersible thermometer aboard in order to determine the temperature of the water before plunging in for a swim. It is useful in many coastal areas, particularly in regions such as New England and California, where warm currents are mixed with cold currents more often than not with one over the other. We dangle our thermometer overboard a few feet below the surface on the end of a long length of marlin and check the temperature before we plunge. Being cowards at heart in this department, toe dabbling has not proved sufficient to save us from the shock of encountering underlying cold, cold water below a sun-warmed surface.

THE SLING PSYCHROMETER

We consider the most important function of a sling psychrometer to be its use in determining dew point with the aid of a dew point table. As previously stated, dew point is the temperature below which the water vapor in the air condenses into dew or fog. As the temperature of the air continues to cool below dew point, some form of precipitation will occur.

A sling psychrometer consists of a frame that holds a pair of mercurial thermometers, one of which is referred to as the dry bulb thermometer and the other as the wet bulb thermometer. They are both like common household thermometers except that the bulb of one is covered with a muslin or heavy gauze sleeve.

The thermometers are suspended side by side from their frame with the bulbs down and are separated from each other enough so that there is plenty of room for ventilation around each of the bulbs.

The temperature of the surrounding air is registered on both thermometers until the bulb wrapped in its gauze sleeve is dipped into clean, fresh water and the entire instrument is swung around through the air by its handle. The measurement of air temperature by the dry bulb thermometer is not affected. However, the swinging motion forces air around the wet sleeve causing the water to evaporate, which cools the bulb on the wet bulb thermometer; the mercury falls to register a temperature lower than the surrounding air temperature. The rate of evaporation, as demonstrated by the amount the temperature has dropped on the wet bulb thermometer, is dependent on the amount of water vapor currently present in the air.

Repeat the same procedure a few times to be sure of a realistic reading and then subtract the wet bulb degrees from those of the dry bulb. The difference is referred to as the depression. The dew point temperature is then determined by entering a dew point table with the depression and the air temperature figures.

We include a portion of the dew point table because it is usually difficult to find unless you have access to certain physics or chemistry textbooks or to the U.S. government's Pub. No. 9, Volume II, *American Practical Navigator* (Bowditch), where a table covering a greater range may be found.

With a psychrometer, you can make a comparison of a series of dry bulb and wet bulb air temperatures, which is often useful for forecasting condensation or precipitation in your local area. For example, you can expect the possibility of fog or rain when the spread between the two temperatures is narrowing. The spread narrows when either the air becomes colder and its temperature drops to approach the dew point or moist air moves into an area and the dew point rises to approach the air temperature.

When air temperature levels off before it reaches the dew point, as it does frequently, condensation does not occur. Expect clear weather when there is a large spread between the two measurements or if the spread is increasing.

Your local dew point can be compared with those reported by aviation and marine weather stations to the west of your location as an additional aid to making your own predictions.

The sling psychrometers we have seen include only the table for determining relative humidity. This measurement of the amount of moisture in the air is frequently given on radio and television weather reports intended for the general public. The relative humidity is the ratio of the amount of water vapor that a body of air can hold at the current temperature, and it is usually expressed as a percentage. As the air temperature decreases, the relative humidity increases; condensation occurs when the relative humidity is 100 percent. Air temperature and the depression are used to enter a relative humidity table. This table can be found in many scientific textbooks as well as in the *American Practical Navigator*.

DEW POINT TEMPERATURE TABLE

The dew point is the air temperature below which condensation will occur in some form. To use this table, enter it down the left hand column with the dry bulb air temperature degrees and across the top with the depression. The degrees of the dew point temperature can be read at the intersection of the two.

Dry Bulb Air Temp. (°F)	Depression (F)									
	1°	2°	3°	4°	5°	6°	7°	8°	9°	10°
+40	+38	+35	+33	+30	+27	+24	+20	+16	+11	+ 4
42	40	38	35	33	30	27	23	19	15	10
44	42	40	37	35	32	29	26	23	19	14
46	44	42	40	37	35	32	29	26	22	18
48	46	44	42	40	37	35	32	29	26	22
50	48	46	44	42	40	37	35	32	29	25
52	50	48	46	44	42	40	37	35	32	29
54	52	50	49	47	44	42	40	37	35	32
56	54	53	51	49	47	45	42	40	37	35
58	56	55	53	51	49	47	45	43	40	38
60	58	57	55	53	51	49	47	45	43	40
62	60	59	57	55	54	52	50	48	45	43
64	62	61	59	57	56	54	52	50	48	46
66	64	63	61	60	58	56	54	52	50	48
68	67	65	63	62	60	58	57	55	53	51
70	69	67	66	64	62	61	59	57	55	53
72	71	69	68	66	64	63	61	59	58	56
74	73	71	70	68	67	65	63	62	60	58
76	75	73	72	70	69	67	66	64	62	61
78	77	75	74	72	71	69	68	66	65	63
80	79	77	76	74	73	72	70	68	67	65
82	81	79	78	77	75	74	72	71	69	67
84	83	81	80	79	77	76	74	73	71	70
86	85	83	82	81	79	78	76	75	74	72
88	87	85	84	83	81	80	79	77	76	74
90	89	87	86	85	84	82	81	79	78	76
92	91	89	88	87	86	84	83	82	80	79
94	93	92	90	89	88	86	85	84	82	81
96	95	94	92	91	90	88	87	86	84	83
98	97	96	94	93	92	91	89	88	87	85
100	99	98	96	95	94	93	91	90	89	87

Fig. 9. DEW POINT TEMPERATURE *The National Weather Service advises that it will probably continue to determine the dew point in degrees Fahrenheit because it is more precise and the dew point temperature figure will be converted to degrees Celsius for reporting.*

The Celsius-Fahrenheit conversion tables may be found in Appendix B.

You can easily make your own wet bulb/dry bulb instrument that will achieve the same results as a commercially marketed sling psychrometer for a smaller investment. We purchased two household thermometers and fastened a gauze pad from the medicine chest around the bulb of one of the thermometers with a rubber band. The wet bulb reading is taken after the gauze is soaked with fresh water and this thermometer is either swung through the air on the end of a string attached to the top or vigorously fanned with a magazine or notebook. The rest of the procedure is just the same as described above. Incidentally, the wet bulb of a sling psychrometer can be fanned as well, instead of swung about, when swinging room is restricted.

CHAPTER 4

Behold the Super Weather Sleuths

By now your eyes have been opened to the various ways you can match your weather wits at a local level against those of the trained meteorologists who make their predictions available to you daily through various media.

It's a real feather in your cap when you are right about your local conditions and the professionals are wrong; and this sometimes happens. However, it just may be that the few times you come out on top in this detecting and predicting game, it is because of how little you know and not how much. The National Weather Service is accumulating, assimilating, and disseminating vast amounts of weather information daily, and in their forecasts they continually make judgments on far-reaching weather probabilities you cannot even consider.

If local conditions sometimes catch them unawares, be charitable; after all, Mother Nature has no intention, at least to date, of yielding up her secrets easily.

DATA ACCUMULATION

This is an appropriate time in your study to delve into the resources available to meteorologists in order to realize the scope of weather investigation.

Thousands of land-based manned and automatic weather observation stations collect earth surface information in this country. A pamphlet published by NOAA states that of the approximately 1,000 staffed surface observing stations within the United States, only about

300 are directly run by the National Weather Service. Others are administered by the U.S. Coast Guard, the U.S. Navy, the Federal Aviation Agency, and other governmental agencies as well as by private citizens and privately owned companies in cooperation with the National Weather Service.

Surface stations responsible for the extensive weather information necessary for aircraft operation usually transmit weather information every hour—and more frequently if there are significant weather changes. Automatic stations record instrument readings such as temperature, dew point, wind speed and direction, and barometric pressure. At the manned stations, visual observations like cloud cover, visibility, and types of weather and precipitation supplement the instrument readings.

On the high seas, observers aboard several thousand merchant ships in both U.S. and foreign registries act as weather reporters within the Cooperative Ship Program of the World Meteorological Organization (WMO). They are referred to as cooperative observers. When necessary, these cooperative observers are trained by the National Weather Service and may even be provided with meteorological instruments. The Great Lakes Observing Program has cooperative observers aboard car ferries and coal, ore, and grain carriers. Reports from all these commercial ships are made regularly at six-hour intervals at the world Standard Synoptic Times of 0000, 0600, 1200, and 1800 Greenwich Mean Time. (Greenwich Mean Time is discussed later in this chapter.)

According to United States Coast Guard instructions, "Weather reports shall be made by all Coast Guard cutters that have a radioman on board and capacity for either radiotelegraph or teletypewriter when at sea." New instructions try to encourage all ships to report in coastal waters also as winds even near shore sometimes differ markedly from observations at shore stations. U.S. Navy ships in passage record weather conditions in their ships' logbooks and give special weather reports if requested by the National Weather Service when they are in areas of severe storms, hurricanes, or waterspouts.

Monster buoys, as they are referred to by the National Weather Service, are gigantic buoys located in offshore waters to record and transmit weather data. They are 10 to 12 meters (33 to 40 feet) in diameter and weigh up to 91 metric tons (100 tons). The few in operation at present are moored about 90 to 175 nautical miles off U.S. seacoasts.

All the buoys are equipped with instruments that monitor such weather clues as precipitation rate, air temperature and pressure, dew point, wind speed and direction, surface water temperature, and wave height and period. These behemoths of the sea check all their sensors once an hour and transmit their stored information, usually every three hours, to the U.S. Coast Guard radio station in Miami or San Francisco. During critical periods of unsettled weather, the buoys can be questioned once each hour.

They have proved extremely valuable in locating the formation or intensification of many ocean storms. More of the buoys will be added to our waters when funding permits. Max Mull, recent chief of Marine Weather Services, says, "Hopefully, there will be a picket line of ocean buoys around the U.S. coasts."

High in the air over land and sea, governmental, commercial, and private aircraft are in frequent radio communication with ground stations, and pilots transmit their weather observations as part of these routine communications. In addition, radar is used to scan the air for the extent, intensity, and movement of areas of precipitation, severe thunderstorms, tornadoes, and hurricanes; balloons record weather data from the surface of the earth up to about 30 000 meters (100,000 feet); rockets collect information between about 30 000 to 1 000 000 meters (100,000 to 300,000 feet); and satellites, which remain stationary in space in relationship to the earth below them, relay conditions on earth and aloft from their vantage points above the earth's surface.

The satellite program is one of the major contributions of the United States to the World Weather Program. It is hoped that this system of celestial observers will soon be increased to provide a continuous view of the surface of the earth. The World Weather Program is composed primarily of the Global Atmospheric Research Program, which promotes technological development, and the World Weather Watch, which aims to increase the accuracy of weather predictions, especially on a long-range basis.

Satellite contributions to the gathering of weather data are demonstrated on television weather reports when their photographs are used to show the movement of weather systems. Residents along the Atlantic and Gulf of Mexico coasts are interested during hurricane season in the location of any whirling masses of tropical storm clouds.

"Since the beginning of the operational satellite system," an article in the magazine *NOAA* states, "probably no tropical storm at sea has gone undetected. Countless thousands of lives have been spared in the United States and abroad because of the early detection and continued tracking of hurricane and tropical storms."

LONG-RANGE PREDICTIONS

Weather outlooks of one to three months are made by the National Weather Service's Long Range Prediction Group from data compiled over the last twenty-five years of weather registered within the troposphere, the layer of the atmosphere that extends up to 11 kilometers (8 miles) above the earth. Outlooks usually estimate whether the temperature will be above or below normal and sometimes they are expanded to precipitation expectations. These outlooks are computed from changes in upper air patterns.

Many of the predictions result from the location and strength of the polar-front jet stream. Speeds in excess of 240 kilometers per hour (150 statute miles per hour) are often recorded in this narrow band of air; it encircles the world about 10 000 to 12 000 meters (35,000 to 40,000 feet) above the earth; and it meanders northward and southward high over the middle latitudes of the Northern Hemisphere. It is a major factor in weather experienced in the United States because it governs the course of the high and low pressure areas as they sweep eastward across the country.

As the National Weather Service is continually sounding the upper air, they can determine the changing path and wind speeds of the jet stream within the Prevailing Westerlies and predict the general consequences of these changes on weather in the United States with some degree of accuracy.

In the fall of 1976, for example, the Long Range Prediction Group forecast unusually cold weather for the forthcoming winter. As projected, early in 1977 record-breaking cold temperatures were experienced in the midwestern, eastern, and middle southern and southeastern portions of the United States. The clue to this type of weather was a southward shift of the jet stream. This southerly shift had been observed the previous summer; it resulted in the movement of all but one of the hurricanes of 1976 along a predictable track up the middle of the Atlantic Ocean and away from U.S. coastlines.

NATIONAL WEATHER WARNINGS

Four categories of weather warnings are used by the National Weather Service to alert mariners in coastal waters of impending weather conditions. These warnings are reported in newspapers, on television, by radio, and possibly by visual flag displays (see page 35). The four classifications are defined by the National Weather Service as follows:

Small Craft Advisory (used only in coastal waters): to alert mariners to sustained (more than two hours) weather or sea conditions, either present or forecast, that might be hazardous to small boats. Mariners learning of a Small Craft Advisory are urged to determine immediately the reason by tuning their radios to the latest marine broadcasts. Decision as to the degree of hazard will be left up to the boatman, based on his experience and size and type of boat. The threshold conditions for the Small Craft Advisory are usually 18 knots of wind (less than 18 knots in some dangerous waters) or hazardous wave conditions.

Gale Warning: to indicate that winds within the range 34 to 47 knots are forecast for the area.

Storm Warning: to indicate that winds 48 knots and above, no matter how high the speed, are forecast for the area. However, if the winds are associated with a tropical cyclone (hurricane), the storm warning indicates that winds with the range 48 to 63 knots are forecast.

Hurricane Warning: issued only in connection with a tropical cyclone (hurricane) to indicate that winds 64 knots and above are forecast for the area.

Note: In bulletins disseminated to the general public, a "hurricane warning" is a warning that one or more of the following dangerous effects of a hurricane are expected in a specified coastal area in 24 hours or less: (a) sustained winds of 74 statute miles per hour (64 knots) or higher; (b) dangerously high water or a combination of dangerously high water and exceptionally high waves, even though the winds expected may be less than hurricane force.

On the other hand, a "hurricane watch" is an announcement that a hurricane, or incipient hurricane condition, poses a threat to coastal and inland communities. All people in the area should then take stock of their preparedness requirements, keep abreast of the latest advisories and bulletins, and be ready to take action if a warning is issued.

Other types of warnings important to boaters follow.

Special Marine Warning Bulletin. This is issued whenever a severe local storm or strong wind of brief duration is imminent and is not covered by existing warnings or advisories. Boaters will be able to receive these special warnings by keeping tuned to a NOAA VHF-FM radio station or to U.S. Coast Guard and commercial radio stations that transmit marine weather information.

Thunderstorm Watch and Warning. These advisories should be heeded as follows: in case of a "thunderstorm watch" keep in touch; in case of a "thunderstorm warning" take protective measures *now*—minutes are important!

MARINE WEATHER SERVICES CHARTS*

A series of *Marine Weather Services Charts* is prepared by the National Weather Service for the coastal boating areas of the United States as well as for the Great Lakes. Some charts have a short résumé of weather conditions likely to be found in the specific areas covered. All of the charts include an explanation of National Weather Warnings.

Listed on these charts are the locations and frequencies of VHF-FM weather broadcasts as well as the times, locations, and frequencies of marine forecasts and warnings broadcast on AM and FM radio and marine radiotelephone stations. This information is also given for aviation weather broadcasts from air navigation radio stations, as well as for storm information broadcasts on the time broadcast radio stations WWV and WWVH. Telephone numbers for the National Weather Service offices in the areas on each chart are given also. All of these different types of weather broadcasts are discussed in Chapter 5.

An outline of the area covered by each *Marine Weather Services Chart,* along with its number, is depicted on Fig. 10 (see page 63).

These charts are available for the following areas and should be ordered by number as well as by location.

MSC-1 Eastport, Maine to Montauk Point, N.Y.
MSC-2 Montauk Point, N.Y. to Manasquan, N.J.
MSC-3 Manasquan, N.J. to Cape Hatteras, N.C.
MSC-4 Cape Hatteras, N.C. to Savannah, Ga.
MSC-5 Savannah, Ga. to Apalachicola, Fla.
MSC-6 Apalachicola, Fla. to Morgan City, La.
MSC-7 Morgan City, La. to Brownsville, Texas
MSC-8 Mexican Border to Point Conception, Calif.
MSC-9 Point St. George, Calif. to Point Conception, Calif.
MSC-10 Point St. George, Calif. to Canadian Border
MSC-11 Great Lakes: Michigan and Superior
MSC-12 Great Lakes: Huron, Erie, and Ontario

---

* Order from: Distribution Division (C44), National Ocean Survey, Riverdale, Maryland 20840.

MSC-13 Hawaiian Waters

MSC-14 Puerto Rico and Virgin Islands

MSC-15 Alaskan Waters

WORLDWIDE MARINE WEATHER BROADCASTS*

NOAA's publication *Worldwide Marine Weather Broadcasts* is intended for all U.S. ships wherever they may be, and it incorporates a great wealth of information on the many and varied transmissions available to transoceanic voyagers. Frequencies and times of radiotelephone weather broadcasts are listed in this annual publication.

Almost all ocean-going ships have personnel aboard skilled in radiotelegraphy who send and receive messages in International Morse Code. Some large vessels even have radio equipment capable of receiving facsimile transmissions of surface analysis and prognosis charts similar to your surface weather chart (Fig. 1). *Worldwide Marine Weather Broadcasts* contains details for these weather transmissions as well as the customary AM and FM radiotelephone broadcasts.

This book is divided into five parts: Section 1, Radiotelegraph Broadcasts; Section 2, Radiotelephone Broadcasts; Section 3, Radiofacsimile Broadcasts; Section 4, Radioteleprinter Broadcasts; and Section 5, Weather Broadcasts for the Great Lakes Continuous Broadcasts—NOAA Weather Radio (VHF-FM). The first four sections are arranged into large ocean zones such as the North Atlantic Ocean, the South Atlantic Ocean, and the like. Under each zone, there is a list of the transmitting stations that give weather broadcasts.

TIMES USED IN REPORTING WEATHER

Some of the ways in which time is expressed in weather reporting may be unfamiliar to you. For example, the hours and minutes of a day are often given in specialized broadcasts, such as those for marine and aviation weather, in four-digit, 24-hour time.

The hours of the day in 24-hour time are identified on a consecutive 24-hour basis instead of the familiar two 12-hour segments associated with watch time. Watch times, as you know, require the label A.M. for the hours before noon and P.M. for the hours after noon. Watch time is used on local weather broadcasts intended for the general public,

* Order from: Superintendent of Documents, U.S. Government Printing Office, Washington, D.C. 20402.

such as those on television and AM and FM radio entertainment media.

When four numbers in 24-hour time are used, the first two digits represent the hours, and the last two, the minutes. The last minute of a day is expressed as 2359, and a new day starts at 0000; 1200 is noon. Therefore, 2:00 A.M. watch time is 0200 in 24-hour time and 2:10 P.M. is 1410. A memory aid for converting watch time to 24-hour time is "after noon, add twelve."

The four digits of 24-hour time are usually spoken separately and thus 1410 would be transmitted as, "one-four-one-zero". Sometimes the verbal expression of whole hours is shortened and 0200 is spoken as "zero-two-hundred" and 1400 as "fourteen-hundred."

Times used by ships at sea to report weather conditions are called World Standard Synoptic Times. These times are 0000, 0600, 1200, and 1800 Greenwich Mean Time (GMT). GMT is also used in broadcasts intended for ships and aircraft with a wide range of travel. You will find it necessary to convert the GMT used to your standard zone time before you will be able to determine the possible effect the weather events as reported in GMT may have on your local area.

A few hundred years ago, the official time in England was reported once a day from the Royal Observatory in Greenwich. Gradually, other nations adopted this time so that it became universal. As time and longitude are interrelated, the original Royal Observatory in Greenwich became the starting point in 1884 for dividing the east-west measurement of the earth's surface into 360 degrees of longitude and into 24 time zones. This starting point is marked today with a bronze strip.

Longitude is expressed in degrees (°) and minutes (′), and is measured from 0° longitude, which is an imaginary line running from the North Pole to the South Pole through the original site of the Royal Observatory. The 360 degrees of longitude are divided in half with long. 180° extending from the North Pole to the South Pole on the opposite side of the globe from Greenwich, England. Longitudes to the east of Greenwich are identified by the letter E and those to the west, with the letter W.

One hour of time is equal to 15° of longitude. Time is measured from Zone 0 which is bisected by long. 0°; long. 7°30′W and long. 7°30′E are the limits of the 15° of this zone. East and west time zones are numbered consecutively from Zone 0 and converge in the middle of Zone 12 at long. 180°. Long. 180° is the International Date Line where you gain one calendar day as you cross it going west and lose a day heading east.

The time used within each zone is referred to as local standard time. (Wherever daylight saving time is in effect within a zone, one hour must be added to the standard time of that zone.) The time within Zone 0 is called Greenwich Mean Time and the numerical designations of 1 through 12 of the other zones indicate the number of hours the standard time of each zone differs from Greenwich Mean Time. A west longitude zone number is added to the standard time of zones to the west of England in order to determine the GMT (time at Greenwich, England, at that moment); either a plus (+) sign precedes these zone numbers or the letter W follows them. East longitude zone numbers are subtracted from standard zone times to find GMT; either a minus (−) sign precedes these zone numbers or the letter E follows them.

Once you associate the time difference between your standard zone time and GMT with the earth's 24-hour rotation relative to the position of the sun, the time difference between your zone and Greenwich, England, becomes logical. Thus, when the sun rises in Greenwich at 0600, you will not see it rise over the horizon at 0600 in Zone +5 on the U.S. Atlantic coast until five hours later. When the sun rises at 0600 standard time in Zone +5, it will be 1100 GMT (five hours later in the day in Zone 0, Greenwich, England); three hours from this time it will be 0600 on the Pacific coast in Zone +8 or 1400 GMT.

As all the time zones for the United States are entirely to the west of Greenwich, all U.S. zone numbers are added to local standard time to determine the time in Greenwich, England. However, in order to have weather information using GMT meaningful to your area, you will be concerned with making the conversion from GMT to your zone time. To do this, you must subtract any west zone hours from the GMT given in a broadcast.

Time zones in the United States are more commonly known by the following names instead of by zone numbers:

Zone +5 —Eastern Standard Time (EST)
Zone +6 —Central Standard Time (CST)
Zone +7 —Mountain Standard Time (MST)
Zone +8 —Pacific Standard Time (PST)
Zone +9 —Yukon Standard Time (YST)
Zone +10—Alaska-Hawaii Standard Time (AST)
Zone +11—Bering Standard Time (BST)

Fig. 10 depicts the time zones for the United States. The drawing also shows that, although the convention of 15° of longitude within each time zone is generally followed for ocean areas, there is considerable arbitrary variance in this on land areas. Some zone demarcations diverge from longitude lines to make a uniform time possible everywhere within a state. As you can also see from the figure, the International Date Line does not strictly follow long. 180°; instead it has been relocated for practicality to fall between countries.

To further complicate the universal time language, the U.S. National Bureau of Standards changed the name of Greenwich Mean Time (GMT) to Coordinated Universal Time (UTC) on their time signal radio broadcasts. As you will note, the bureau has also changed the order of the initials in the abbreviation of the new name. We were informed that the name alone has been changed, the time expressed remains the same. These time signal broadcasts also include brief weather summaries, which are discussed in Chapter 5.

To sum up, many methods of expressing one specific time on earth can be used. To illustrate this, all of the following are for the same minute in time:

1:00 A.M. —watch time for standard time in Zone +5

0100 A.M. —24-hour standard time in Zone +5

2:00 A.M. —daylight saving time (DST) in Zone +5

0200 A.M. —24-hour daylight saving time in Zone +5

6:00 A.M. —watch time for standard time in Greenwich, England

0600 A.M. —GMT

Note: Whenever the zone difference applied to GMT results in less than 0000 hours or more than 2400 hours, there will be a change in date as well as hours. Assume it is June 7 at 2200 or 10:00 P.M. in the U.S. Pacific coast Zone +8. It will be 0600 GMT or 6:00 A.M. on June 8 in England (2200 hours plus 0800 hours is 3000, which, in computing time, is one day of 2400 hours with an additional 0600 hours on the new date).

Fig. 10. USA TIME ZONES AND MARINE WEATHER SERVICES CHART COVERAGE *The USA time zones were determined from the Standard Times Chart of the World published by the U.S. Naval Oceanographic Office. Time zones consist of 15° of longitude each, except where they are altered for convenience of land areas as is indicated by dashed lines.*

Outlines of the MSC (Marine Weather Services Charts) with their numbers show the areas covered by each of these charts.

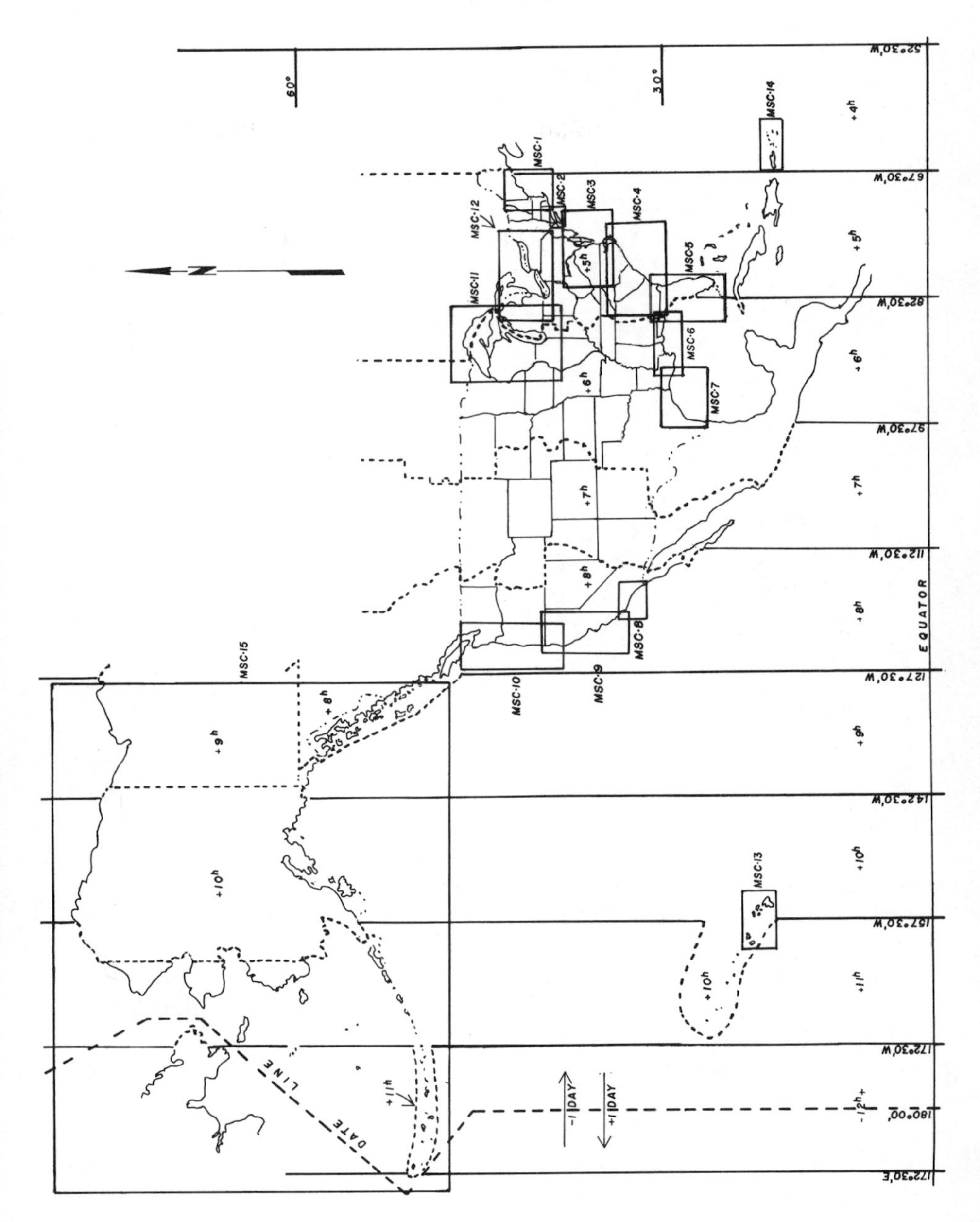
60°
30°
MSC-1
MSC-2
MSC-3
MSC-4
MSC-5
MSC-6
MSC-7
MSC-8
MSC-9
MSC-10
MSC-11
MSC-12
MSC-13
MSC-14
MSC-15
52°30'W
67°30'W
82°30'W
97°30'W
112°30'W
127°30'W
142°30'W
157°30'W
172°30'W
180°00'
172°30'E
EQUATOR
DATE LINE
-1 DAY
+1 DAY
+4h
+5h
+6h
+7h
+8h
+9h
+10h
+11h
-12h+

CHAPTER 5

Utilize Weather Broadcasts

Safety-minded boating enthusiasts, and in particular those with cruising, sailboat racing, and offshore fishing experience, make listening to a variety of official weather broadcasts a part of their everyday lives. One sailing wife says, "Listening to radio weather broadcasts is like putting on your Top-Siders,* you do it automatically; it's all safety."

RADIO COMMUNICATION TERMS

A special language is used for technical discussions of radio communication, which is likely to be frustrating without some clarification.

Radio is a form of telecommunication for transmitting information through the air by electromagnetic waves. Telecommunication is defined by the Federal Communications Commission as, "Any transmission, emission, or reception of signs, signals, writing, images, and sounds or intelligence of any nature by wire, radio, optical, or other electromagnetic systems." The Federal Communications Commission (FCC) is the U.S. Government agency that regulates communications and radio transmissions and issues licenses for radio stations.

To further the understanding of the novice to this field and to promote maximum use of radio for weather broadcasts as discussed later in this chapter, we include, in as nontechnical a presentation as we can give, some of the terms used in radio communication. We suggest that reference to Fig. 11, Radio Waves, will make the explanation of the terms easier to understand.

---

* Registered trademark for Sperry Top-Sider nonskid boat shoes.

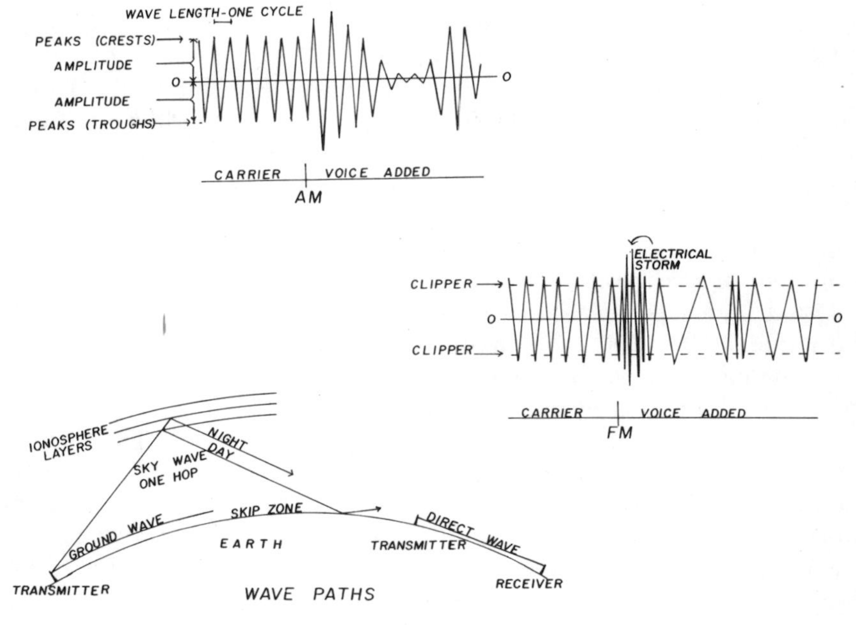

Fig. 11. RADIO WAVES

Intelligence: information conveyed by a communication medium, including telegraphic signals as well as voice transmissions.

Amplitude: the high points of the crests of a radio wave and the low points of its troughs relative to zero, which is the middle point between the two extremes.

Cycle: one complete radio wave sequence as from crest to crest or trough to trough.

Frequency: the number of times per second a cycle is completed. Units of radio frequency are expressed in the following two ways:

KiloHertz (kHz): 1,000 cycles per second. (The prefix *kilo* means one thousand; in the suffix *Hertz,* one Hertz = one cycle per second and is named for a famed German physicist.)

MegaHertz (MHz): 1,000,000 cycles per second. (The prefix *mega* means one million.)

For convenience, in the higher frequencies kiloHertz are converted to megaHertz (1 MHz = 1,000 kHz).

Radio frequencies are used for many different transmissions, including entertainment radio, television, citizen's band radio, electronic navigation, and radiotelephone ship-to-ship and ship-to-coast communication. Frequencies for one type of broadcasting service are grouped together in a band. Specific frequencies are assigned by the FCC and may be changed from time to time.

Ground Waves: radio waves that travel parallel to the earth's surface from the transmitter. The paths of these waves tend to curve around the earth's surface at low frequencies and, when unobstructed, the waves can cover long distances before fading out. The power of a transmitting station is instrumental in determining the distance these radio waves travel.

Sky Waves: radio waves that are reflected back to earth by ionized layers of the atmosphere. These layers can be 140 to 240 kilometers (90 to 150 miles) above the earth's surface and are referred to as the ionosphere. Sky waves are used for long-distance communications and the power of the transmitter is important.

Frequency selection may vary with the time of day. The makeup of the ionosphere changes at night and it is located farther from the earth. When the sun no longer influences the ions, some layers of the ionosphere become less dense and others disappear entirely. During the day, the ionosphere tends to absorb low frequencies and to reflect high frequencies back to earth. At night, the high frequencies often penetrate the ionosphere and the low frequencies are reflected. Our communication's expert at NOAA sums up frequency selection as follows: "As a rule of thumb: day go high, night go low."

Sometimes a sky wave makes one or two hops in its excursions from the earth to the ionosphere and back. The ionosphere is less effective in reflecting radio waves over about 30 MHz.

Skip Zones: areas on earth that lie between the limits of ground wave reception and sky wave reception where a radio signal is not received. At night skip zone areas increase because the transmission path of a radio signal fluctuates with the changes in the ionosphere.

Carrier Wave: the basic radio wave of constant frequency that is assigned to the station. At constant frequency and amplitude this wave contains no information. Information can be superimposed on it at the transmitter station by modulating (changing) its frequency or amplitude as described below.

Amplitude Modulation (AM): a change in the amplitude of a carrier wave by information that is superimposed on it. The carrier frequency remains constant.

Frequency Modulation (FM): information is superimposed on the carrier wave by changing its frequency. The amplitude peaks of the carrier wave are cut off at a specified height in FM receivers by a clipper circuit; annoying interference is greatly reduced as the receiver is only sensitive to modulated frequency rather than amplitude.

(Atmospheric interference tends to amplitude modulate the carrier with noise.) With FM, the carrier wave stretches out and contracts like the bellows of an accordian when information is superimposed on it.

The FCC restricts FM to a higher frequency range than AM. At the high frequencies assigned to FM use, ground waves do not curve. Instead, they travel in a direct line from the transmitting antenna to the receiving antenna, thus reducing the range of FM to the line-of-sight distance. This limitation, however, allows stations only a couple of hundred kilometers apart to use the same frequency without interference. This is the reason that the National Weather Service needs only three frequencies for their series of continuous weather broadcast stations located intermittently along U.S. shores and in the interior sections of the country as well.

Interference: any electromagnetic influence that affects the transmitted signal undesirably. Common sources of interference include atmospheric disturbances such as lightning, engine ignition circuits, and other radio signals. As most of these disturbances affect the amplitude more than the frequency of the carrier, AM transmissions are much more prone to static than FM transmissions.

AM AND FM RADIO ENTERTAINMENT BROADCASTS

The flick of a knob will turn on the entertainment band of an AM or FM radio in almost every home, car, or boat in this country. If a boat does not have a permanently installed radio with an antenna, it is easy and convenient to carry a small battery-operated set aboard. Many AM and FM radio stations broadcast marine forecasts and warnings on a regular schedule. The frequencies and locations of these stations are listed in the *Marine Weather Services Charts.*

You can also learn quite a bit from the brief weather information included in most hour and half-hour radio news programs even though they are not the standard marine weather reports. For example, assume it's a Friday night and you have invited your golfing boss and his wife to join you Saturday aboard your boat. They accept with delight, mentioning happily that they have never been aboard a boat before. As this will be their first experience, you begin to feel a bit apprehensive about the success of the venture because you have heard a mild cold front is on the way. Best to step outside and observe the local scene; it's still warm from the high temperatures of the day and only a gentle wind is rustling the leaves in the trees.

At bedtime you turn on the radio to a local news broadcast for a last weather check and hear the weather report as follows: "There is a possibility of early morning thunder showers, but the rain will be over by daybreak. Tomorrow will be sunny and it will be considerably colder than today."

Now you can relax in peaceful slumber, knowing the front should go through well before it is time to go out on the boat. It will be colder than you had hoped, but the day should be pleasant and crystal clear without possibly unsettling rough seas. There should be only light winds, for it did not blow hard ahead of the front, and newspaper and television reports earlier gave no indication that the front was a strong one. All you have to remember is to call your guests in the morning to suggest they bring along a couple of extra sweaters.

Many radio sets are available with both AM and FM bands. As you know, AM stations have a longer range than FM but FM reception is relatively free from noise and interference. The drawback of static on AM stations, however, can be turned to advantage by weather-minded boatmen.

This is attested to by Proctor Brady, a friend of the Townsends, who initially began cruising in his powerboat many years ago when weather broadcasts were often sketchy or nonexistent. He discovered that static on his AM radio was the best indication he had for judging the severity of any squalls that might be coming his way.

He off-tuned his AM radio to a space near but not on the stations in the lowest range of his radio, between 540 and 600 kHz. He turned up the volume to about two-thirds capacity, just low enough so the static did not blast him out of the boat but high enough so he could hear it, if it was present.

"When there is short staccato static," Proc explains, "the storms are likely to be scattered and not too severe, but if there is a continuous crescendo there is probably a line squall on the way and it is best to stay in port. If the intensity of the static is low and it sort of whispers at you, the weather system is probably two or three hours distant but if it is quite strong and loud a storm is likely to approach within the hour." Even with the fine marine weather reporting today, he still uses this method as an additional check on the weather.

VHF-FM NOAA CONTINUOUS WEATHER RADIO BROADCASTS

A large section of the boating public relies, with good reason, on the VHF-FM continuous marine weather radio broadcasts prepared by the National Weather Service. Weather information is immediately available on one of the three specially assigned frequencies and usually is revised every three hours, with the additional plus that FM reception is almost static-free. Generally, transmitting stations along the shores of boating waters alternatively use the frequencies 162.40 MHz, 162.475 MHz, and 162.55 MHz to broadcast reports and forecasts continuously, 24 hours a day, on taped messages repeated every four to six minutes.

The frequencies and locations of the stations are printed on the *Marine Weather Services Charts* for each area and some of these charts also depict the expected radius of reception for the transmissions. A note on some of these charts states:

> The frequencies 162.55, 162.475, and 162.40 MHz require narrow band FM receivers of ±5 kiloHertz deviation. In selecting a suitable receiver, special attention should be paid to the manufacturer's rating of the receiver's sensitivity. Generally speaking, a receiver with a sensitivity of 1 microvolt or better and a quieting factor of 20 db. (decibels) should pick up a broadcast at a distance of about 40–50 nautical miles depending upon antenna height and terrain.

Frequency information can also be found in *Worldwide Marine Weather Broadcasts*.

Radio stations for the VHF-FM continuous weather broadcasts are managed by the National Weather Service. Forecasts are issued every six hours; broadcast tapes are updated frequently, and amended as required. Contents of the broadcasts vary, but in general contain the following types of information:

1. Description of the weather patterns affecting the region
2. Forecasts and warnings for nearby coastal waters or appropriate Great Lakes
3. Local weather observations and forecasts
4. Radar summaries and reports
5. Special bulletins and statements concerning severe weather or hurricanes
6. Weather observations from selected National Weather Service and Coast Guard stations
7. Harbor and river water levels

As VHF-FM broadcasts are line-of-sight communications, their range is dependent on the height of the antenna of the transmitting stations; the type and height of the antenna on your set; and whether or not there are any intervening land masses, buildings, or other solid obstacles between your radio and the broadcasting station antenna. VHF-FM transmissions usually can be received 32 to 64 kilometers

(20 to 40 statute miles) from the sending antenna and upon some occasions more than 160 kilometers (100 statute miles) from it.

There are several compact, reasonably priced transistor radios on the market today, some of which are designed primarily for receiving the continuous weather broadcasts. In addition, there are large multiband radio sets that also can receive these broadcasts. The frequency range of the continuous weather stations falls within the Public Service Band of 147 MHz to 174 MHz, which also includes police and land mobile service.

We have noticed while cruising along the U.S. coasts that even aboard auxiliaries equipped with mast-top antennas there are still a few gaps where it is impossible to receive the continuous weather broadcasts. The National Weather Service plans to upgrade the system with new recording equipment, improved antennas, and antenna locations. The criterion will be for all stations to have transmitting capabilities of at least 64 kilometers (40 statute miles).

Some stations omit so much material in their reports that we feel we are not being adequately informed of weather conditions. For example, on a recent east coast cruise we found that the weather service in Charleston, S.C., gave one of the most complete and thereby most useful broadcasts on the east coast. It included a discussion of the large overall picture of the locations of fronts in the United States and Canada as well as local reports and forecasts for Charleston Harbor, the South Carolina Lakes, and inland South Carolina. It also gave reports for North Carolina, South Carolina, Georgia, and Florida coasts up to 17 nautical miles offshore. These reports included tides, beach erosion, wave heights, and radar summaries. In addition, the tropical weather outlook during hurricane season was discussed including encouraging reports when there were no disturbances. In contrast to this complete and much-appreciated coverage, some of the nearby stations omitted reporting on the location of pertinent North American frontal systems, coastal conditions along the nearby states, and the lack of hurricane activity in the height of the season.

TELEVISION WEATHER BROADCASTS

As pointed out in the discussion of television weather maps in Chapter 1, regularly scheduled TV weather programs are an excellent source for receiving up-to-date weather analysis because they usually report the very latest local and national conditions.

In addition to regular programming, instant weather advisories are usually available for persons with cable TV service. Most cable TV companies reserve one channel for continuous weather information; some broadcast the actual voice transmission of the VHF-FM NOAA Continuous Weather radio broadcasts, while others have a printout of the words of the broadcast on the television screen.

WARNING ALARM

A warning alarm is a system comprised of an automatic device that can be added to a radio or radiotelephone (discussed later in this chapter) or can be built into it. It is activated by the National Weather Service "in the event of sudden weather deterioration such as a severe thunderstorm watch or warning, a passage of a violent squall line, or similar phenomena." These specially designed warning receivers will either sound a shrill alarm indicating that an emergency exists and the listener should tune in the appropriate frequency or, when operated in a muted mode, will be automatically turned on so that the warning message will be heard.

The warning alarm device can be used on VHF-FM sets for the three weather station frequencies of 162.40 MHz, 162.475 MHz, and 162.55 MHz, and on the distress and calling frequency 156.80 MHz, either on home radios or on radiotelephones aboard boats. On radiotelephones, which are used for offshore boating, the warning alarm is activated on the International Calling and Distress frequency of 2182 kHz. The FCC *Rules and Regulations** state that "the purpose of this special signal is the actuation of automatic devices giving the alarm to attract the attention of the operator when there is no one on listening watch on the distress frequency."

AVIATION WEATHER REPORTS

Aviation weather reports are one of the most useful of the many kinds of weather communications because they are revised hourly. You can receive these broadcasts if your radio covers the low frequency beacon band of 200 kHz to 415 kHz. Marine radiobeacon signals are allotted a part of this beacon band by the FCC, making it possible for aviation broadcasts to be received with marine radiodirection finders. The *Marine Weather Services Charts* list weather broadcasts by air navigation radio stations with their frequencies and schedules.

---

* Order from: Volume IV, Part 83, Superintendent of Documents, U.S. Government Printing Office, Washington, D.C. 20402.

A selection from the aviation weather station reports submitted to the National Weather Service each hour is taped on a regional basis for aviation broadcasts by the Federal Aviation Administration (FAA), primarily for the benefit of pilots of small planes. Usually the major airport of an area will transmit 15 to 20 station observations in a broadcast. These are referred to as airway sequence observations. The observations include approximately the same weather information depicted on the specimen station model (Fig. 1) and explained in Chapter 1. In addition to the hourly sequence observations, aviation weather broadcasts include 12-hour area forecasts, revised four times a day, in which general weather systems likely to affect regional weather conditions are reported. Some of the stations broadcast continuously, while others transmit once or twice an hour.

As an example of the station coverage for one region, the sequence for the northeast part of the country, originating from the Atlantic coast city of Boston, Mass., usually contains reports from airports as far west as Syracuse, N.Y., as far south as Philadelphia, Pa., and as far north as Bangor, Maine.

By recording at least three successive hourly aviation broadcasts you can track the dramatic progress, for example, of a strong cold front as it moves from one reporting station to another. As you know, the passage of this type of system is usually shown by a rapid drop of air temperature, a sudden rise in barometric pressure, and a wind shift to a general northerly direction.

When you use a road map to note the locations of reporting stations and to measure the distances between them, you can assess the speed of advance of weather systems while determining their severity from your sequential recordings of the aviation broadcasts. When the weather is to the west of you, you can then determine its probable arrival time in your area.

RECORDING AIRWAY SEQUENCE OBSERVATIONS

Airway sequence observations are always presented in the same format, and with a little practice you will be able to record the information the first time you hear it. The system we have developed to record the messages includes the use of some standard abbreviations as well as some of our own.

First we prepare a lined sheet of paper by writing abbreviations for the conditions given in the reports across the top of the page. Entries are made in the order in which they are transmitted: the stations are listed down the left-hand side of the page and the data for each station is entered next to it under the appropriate headings.

The Boston airway sequence observation portion of an actual broadcast of an aviation weather report originating in Boston, Mass., is quoted below followed by the method we use to record the pertinent information. The rest of the broadcast, which is not quoted, consisted of station observations from other airports. (U.S. customary measurement units of feet, statute miles, Fahrenheit degrees, and inches of mercury were used in this broadcast; to determine how metric units of meters, kilometers, Celsius degrees, and millibars or kiloPascals are expressed, see the Appendix.)

> Following are the weather reports observed at one eight zero zero Eastern Standard Time. Boston; Boston measured ceiling three thousand four hundred overcast, visibility two and one half, moderate rain, fog, temperature five two, dew point five zero, wind one two zero degrees two three, peak gusts two seven, altimeter two niner five three, pressure falling rapidly . . . .

1800 EST

ID	Ceil	Sky	Vis	Wea	Temp/Dew	Wind/Kns	Alt	Rem
BOS	34	Ov	2½	RF	52/50	120/23+27	29.53	PFR

EXPLANATION OF HEADINGS, ABBREVIATIONS, AND CODE FOR RECORDING AN AIRWAY SEQUENCE OBSERVATION

ID: a three-letter abbreviation to identify the city from which the weather observation originates.

Ceil: for ceiling or height measurement figures of the lowest layer of cloud covering more than half of the sky.

Sky: for sky cloud conditions abbreviated as follows: Cl(clear); Sc(scattered); Br(broken); Ov(overcast); and Ob(obscured). The sign "-"(thin or partially) is used for either modifier and precedes the sky conditions, such as -Br(thin broken) or -Ob(partially obscured).

Vis: for visibility distance figures. (Statute miles and fractions of miles were used in this broadcast.)

Wea: for weather conditions signified by a letter, such as R(rain); L(drizzle); S(snow); T(thunderstorm); A(hail); F(fog); H(haze); and K(smoke).

Other letters are used to modify these conditions, such as rain and snow followed by W(showers) would be RW or SW; drizzle or rain preceded by Z(freezing) would be ZL or ZR; fog by G(ground), GF; and snow by B(blowing), BS.

Signs for intensity follow the weather condition letter, such as: "-"(light); "--"(very light); and "+"(heavy); as in "R-"(light rain). Moderate intensity has no sign.

Temp/Dew: for temperature and dew point figures, which are separated with a slash.

Wind/kns: for wind direction given in true degrees and wind speed in knots, which are separated with a slash. Zero is calm. Peak gusts preceded by a "+" sign follow the entry.

Alt: for altimeter setting (for airplanes), which is equal to the barometric pressure at sea level given, in this broadcast in inches of mercury; the first two digits are inches and the second two, hundredths. The altimeter relationship to barometric pressure is explained on page 43.

Rem: for remarks such as "pressure falling rapidly" and "thunderstorms in the northeast," abbreviated as PFR and T NE.

The hour the taped message was made is announced just prior to the sequence observations and is recorded on the lined sheet above the headings.

Sometimes when the aviation weather broadcasts are continuous, there is a delay for a few minutes after the hour while the tapes are being changed. One complete aviation weather report is typically about eight minutes long.

TIME SIGNAL/WEATHER RADIO BROADCASTS

The National Burau of Standards, in cooperation with the National Weather Service, provides brief storm information for mariners on two stations that broadcast precise Coordinated Universal Time (UTC) signals. One station is located in Colorado and the other in Hawaii.

The time of day is announced on the UTC broadcasts at one-minute intervals with audible pulses for the intervening seconds. Three voice announcements of 45 seconds each replace the audible pulses on the time signal to provide high seas storm information every hour. According to a pamphlet by NOAA:

> The broadcast is a brief summary describing the location and movement of low pressure centers that are causing (or are ex-

pected to cause) wind speeds of 34 knots or more. Speed of movement is always in knots although the word "knots" may be omitted for the sake of brevity. The summary also describes areas of maximum wind speed and wave heights associated with each storm center. It further describes areas of lesser wind speed (as low as gale force, 34 knots) with respect to distance and direction from the storm center.

If there are no warnings, this happy news is broadcast instead.

The National Bureau of Standards publishes the following particulars of the storm broadcasts on their two stations:

WWV (Ft. Collins, Colo.) Frequencies: 2.5, 5, 10, and 15 MHz

8 and 9 minutes past the hour, storm information for the western North Atlantic, including the Gulf of Mexico and the Caribbean Sea.

10 minutes past the hour, storm information for the North Pacific east of long. 140°W.

WWVH (Kauai, Hawaii) Frequencies: 2.5, 5, 10, and 15 MHz

48, 49, and 50 minutes past the hour, storm information for the North Pacific, also for the South Pacific to lat. 25°S, long. 160°E to long. 110°W.

Multiple frequencies are allotted by the FCC to each of the stations for better reception. The selection of the best frequency depends on the atmospheric conditions and the time of day. The lower frequencies are usually better at night. You can often receive more than one of the frequencies at any one time on a multiband radio.

The weather information for the broadcasts is provided to the National Weather Service by weather satellites and by the Cooperative Ship Program.

CITIZENS BAND REPORTS

Commercial and private fishermen often communicate with one another by citizens band (CB) radio concerning the weather conditions they are experiencing at their various fishing grounds. CB radios are often reasonably priced and are small enough to be carried aboard easily. Before investing in one of these radios, however, it is best to talk with local fishermen to find out the channel or channels most likely to be used by them in order to take advantage of their weather chitchat.

The year 1978 marks the start of the monitoring of channel 9 by the U.S. Coast Guard for marine emergencies only; weather announcements are not planned on this channel at this time.

NATIONAL WEATHER SERVICE FORECAST OFFICE REPORTS

The National Weather Service, which always has the public's interest at heart, has even taken care of the eventuality that you may not be able to receive their standard marine weather broadcasts: in this event, weather information can be obtained by telephoning the nearest National Weather Service Forecast Office. These are usually located at the airport of a major city. Sometimes a ring-through feature is provided so that by waiting 10 seconds after the recording ends you can ask for additional information.

The telephone numbers of the local National Weather Service Forecast Offices are listed in the telephone book under "United States Government, Commerce Department." The location of offices, the telephone numbers, and other details of the service are given in the *Marine Weather Services Charts.* At sea, away from a land line, the National Weather Service Forecast Offices can be reached on a ship-to-coast frequency aboard boats equipped with a marine radiotelephone.

VHF-FM RADIOTELEPHONE WEATHER BROADCASTS

At sea, a radiotelephone is your only link with other boats and shore other than hearty lung power, a good megaphone, or semaphore signals. Boat owners who wish to have radiotelephones aboard their boats must install very high frequency–frequency modulation (VHF-FM) sets as primary units whether they plan to use the boat for lake, coastal, or offshore cruising.

As stated by the FCC, "Under FCC Rules, VHF-FM capability at a ship station is a prerequisite for licensing in other frequency ranges, and the Rules prohibit the use of other frequencies when within VHF range." The 156–158 MHz portion of the VHF-FM band, which is licensed to recreational ship-to-ship and ship-to-coast use, falls within the 156–162 MHz portion where public coast stations are authorized to transmit by radiotelephone. This is higher than the entertainment portion of the FM band, which is 88 MHz to 108 MHz.

Most owners of VHF-FM sets elect to have the crystals installed for the three receive-only continuous weather frequencies of 162.40,

162.475, and 162.55 MHz in addition to having a portable radio aboard capable of receiving them. Reception is likely to be better on a radiotelephone than on the smaller radios, discussed previously in VHF-FM continuous weather radio broadcasts, because it is more sensitive and has an antenna permanently installed at the highest point on the boat.

The FCC *Rules and Regulations* require each VHF-FM set to include channel 16 (156.8 MHz), the distress, safety, and calling frequency. Whenever the radiotelephone on a pleasure boat is turned on without being used for communication the set must be tuned to channel 16 with the operator maintaining a listening watch. Because of this monitoring requirement, an announcement on channel 16 precedes the broadcast of any special weather warnings to be broadcast on an assigned working frequency or on the Coast Guard channel 22 (157.1 MHz). These same warnings, of course, can also be heard on the VHF-FM continuous weather broadcasts.

In areas where there is no National Weather Service continuous weather broadcast available, the U.S. Coast Guard may broadcast the weather on a regularly scheduled basis. Times of these broadcasts are listed on *Marine Weather Services Charts* for the area and in the *Worldwide Marine Weather Broadcasts* book.

In addition, a few VHF-FM public coast station radio operators regularly broadcast the weather. Two stations to introduce this service were Tampa, Fla., on channel 26 and Mobile, Ala., also on channel 26. You can check with your local marine operator to see if this service is offered in your area.

TWO-MEG BAND SSB WEATHER BROADCASTS

The frequency range licensed by the FCC for one type of single sideband (SSB) radiotelephone transmissions is 1600 kHz to 4000 kHz, and, although the megaHertz conversion to 1.6 MHz through 4 MHz for this range spills over the 2-MHz portion of the frequency band, this frequency span is known familiarly as the 2-meg band.

When a boat is used often more than 20 nautical miles from shore for cruising or fishing and is out of range of VHF-FM transmissions much of the time, the owner may wish to install a 2-meg SSB radiotelephone. It is often possible with a 100-watt SSB set to communicate with boats or shore stations one or two hundred miles away and to receive the marine weather messages given on SSB coast stations.

There are no continuous weather broadcast stations on the 2-meg band. Instead, broadcasts of marine weather by radiotelephone stations are transmitted by the U.S. Coast Guard. Warnings are given by marine operators on public coast stations and some of these broadcast the weather on a regular basis. Times and frequencies for these broadcasts are listed on *Marine Weather Services Charts*. They are also included in *Worldwide Marine Weather Broadcasts*, but, in this publication, they are interspersed with many other frequencies so must be identified either by their A3A or A3J classes of emission or by their call signs.

In place of a radiotelephone, or in addition to it, a great many persons participating in offshore work have a powerful multiband radio receiver aboard. Radiotelephone messages can be received on this radio depending upon the set's sensitivity and frequency capability. Although it requires precise tuning, SSB transmissions can be detected only on a radio that is equipped with a circuit bearing the imposing title of beat frequency oscillator, or as referred to by many in our abbreviated-oriented world, as BFO.

The need for a BFO to pick up 2-meg band transmissions on a radio has not always been necessary because in the past AM transmissions were used in the United States. However, because of the tremendous increase in United States marine radiotelephone traffic in recent years, it was necessary to reduce the space required on the frequency band for each transmission in order to accommodate more stations.

Transmission of intelligence with AM broadcasts requires a bandwidth large enough to accommodate the carrier frequency as well as an upper and lower sideband. Thus, AM radiotelephone transmissions are called double sideband or DSB. Each sideband carries the intelligence, whereas the carrier does not.

In single sideband, all or part of the carrier as well as one sideband is suppressed within the transmitter. SSB uses a smaller bandwidth than DSB and is usually a lot clearer. As described by a marine electronics expert, "SSB has seven times the effective talk power of DSB."

Since the receiver circuits require the presence of the carrier, radiotelephones designed to receive SSB transmissions reinsert it. Messages may also be received on multiband radios by using a BFO circuit to "create" a carrier signal.

Even though the United States has discontinued issuing station licenses for AM radiotelephones (DSB), SSB sets must have some AM capability. By international agreement, 2182 kHz will always be available on DSB radio; 2182 is the 2-meg band international calling and distress frequency, this band's counterpart to VHF-FM's channel 16. SSB sets will either have a switch to change to 2182 AM, or a reinserted carrier.

Any special weather warnings issued on the U.S. Coast Guard frequency of 2670 kHz are preceded by an announcement on 2182 kHz.

COAST GUARD SSB PROGRAM

The Coast Guard has inaugurated a two-way weather information program for SSB HF (high frequency, 3 to 30 MHz) radiotelephones that utilizes meteorological and AMVER reports. AMVER means Automated Mutual Assistance Vessel Rescue and has been operated by the Coast Guard for many years.

In the Coast Guard SSB program, the new wrinkle is that in addition to your being able to contact the Coast Guard for emergencies and listening to the regularly scheduled weather reports, you are encouraged to call in immediately about any unusual weather conditions you experience if you have not heard them reported previously. Certainly many of us can think back on occasions when we have been caught out in just this type of situation listening to a weather report that wasn't even close to the conditions we were experiencing. Now we will be able to do our part in contributing to boating safety.

The areas for the program are Boston, Mass.; Portsmouth, Va.; New Orleans, La.; San Francisco, Calif.; Kodiak, Alaska; Honolulu, Hawaii; and Guam, an island of the Marianas group in the Pacific.

Special crystals installed in SSB sets are necessary to either communicate with the Coast Guard stations or to receive their regularly scheduled weather broadcasts. As in the time signal radio broadcasts, there are several frequencies from which to choose. The frequency range licensed by the FCC for SSB radiotelephone transmissions above the 2-meg band is 4000 kHz to 22,000 kHz, and these frequencies fall within that range.

The following three ship stations and coast stations carrier frequencies give coverage for most of the Atlantic Ocean, the Gulf of Mexico, and the Pacific Ocean for this Coast Guard SSB program:

Ship station carrier frequency

4094.8 kHz

6207.2 kHz

8226.8 kHz

Coast station carrier frequency

4393.4 kHz

6521.8 kHz

8760.8 kHz

(Note: Kodiak only has 8226.8 kHz and 8760.8 kHz capability at this time.)

The class of emission for these SSB broadcasts is A3J. Contact your marine electronics dealer to ask which crystals you should have installed in your set. If more information is needed about this program contact the communications officer by telephone at the nearest Coast Guard district or write: Office of Technical Service (W52), National Weather Service, 8060 13th Street, Silver Spring, Md. 20910.

HIGH SEAS RADIOTELEPHONE WEATHER BROADCASTS

The American Telephone and Telegraph Company operates three public coast stations for long range marine radiotelephone service. This service is called High Seas. There are many frequencies for each station in the 4–23 MHz band for SSB transmissions, and the operator aboard ship selects the best frequency for talking to the shore station or for receiving the regularly scheduled weather broadcasts according to the time of day, the season, and the location of the vessel. The High Seas operator ashore can adjust an antenna to achieve the best communication.

AT&T publishes a booklet entitled *High Seas Maritime Mobile Radio-Telephone Service* that describes the service and lists the times and frequencies of the weather broadcasts. This information is also listed in the radiotelephone section of the *Marine Weather Services Charts* and in the *Worldwide Marine Weather Broadcasts* book by call letter in the proper area.

The booklet may be obtained without charge by writing to the American Telephone and Telegraph Company at one of their three coast stations:

Station WOO, P.O. Box 558, Beach Avenue, Manahawkin, N.J. 08050

Station WOM, 1350 N.W. 40th Avenue, Fort Lauderdale, Fla. 33313

Station KMI, P.O. Box 8, Inverness, Calif. 94937

Our good friends Kirt and Mary Hine, who have cruised extensively in the Atlantic Ocean, the Caribbean Sea, and the Gulf of Mexico, as well as to the Galapagos Islands in the Pacific Ocean, told us about the radio coverage they have aboard their Countess 44 auxiliary ketch. Radio equipment aboard *Marigo* includes a VHF-FM set, a battery-operated multiband radio receiver, and an SSB set capable of receiving six WOM High Seas frequencies. Depending upon atmospheric conditions, Kirt finds that during the day on his six WOM frequencies, the two 4-MHz frequencies provide him a range of 50 to 400 nautical miles; the two 8-MHz, a range of 400 to 1,000 nautical miles; and the two 12-MHz, a range of 1,000 to 1,500 nautical miles. He says the lower frequencies usually provide better reception at night. The daytime frequency range bears out the rough rule given us by a marine electronics expert who said, "Multiply the MHz frequency times 100 to find the range in miles frequencies are usually good for."

With High Seas capability, it is also possible to telephone one of the three coast stations for weather and the operator will play the latest taped message. Ordinarily, High Seas offshore weather reports are very good despite the broad expanses of oceans they cover. However, once when Kirt and Mary were en route to the Virgin Islands from the lower Bahamas, there was a great discrepancy in the wind velocity reported: WOM reported winds of 15 knots gusting to 25 when actually it was 25 gusting to 40. As Kirt says, "Some difference!"

When Kirt called the WOM station at Ft. Lauderdale, Fla., on his radiotelephone to complain, the manager said, "Don't complain to us, we just play the Weather Service tape. I'll transfer you to the service's forecaster in Miami and you can complain to him." The transfer was accomplished and a Miami forecaster told Kirt, "We don't get too many inputs from the area where you are. Would you call us collect every morning between 9:00 and 10:00 A.M. and let us know what you have for weather?"

This Kirt did willingly, although hearing his own weather observations in their location repeated later each day word for word over WOM did not solve his desire for additional weather information.

RADIOTELEGRAPH WEATHER BROADCASTS

Radiotelegraphy is the system of transmitting intelligence by the international Morse code dot-dash system of signals. Most Morse code is transmitted by means of a continuous wave (CW) at a constant frequency that is turned on and off by the operator's key to provide the short and long signals of Morse code. If you do not have a special receiver, you cannot detect CW transmissions. However, any receiver with a BFO within a frequency range of the station can interpret these transmissions when the instrument is precisely tuned. The BFO generates a frequency near the CW carrier and the two "beat" together to produce an audible tone. A BFO, of course, is not necessary when the code is transmitted by amplitude or frequency modulation of a carrier.

We know of several boat owners who make use of telegraphed weather signals even when initially they cannot decipher the code. They receive these coded signals on a multiband radio, record the messages on a two-speed tape recorder, and at the end of the transcription play back the message at a slower speed better suited to an untrained ear. They write down the dot-dash combinations of the code, replaying the tape as necessary and decode the message with a copy of the Morse code translation.

Ships participating in the Cooperative Ship Program, telegraph weather reports at the World Standard Synoptic Times of 0000, 0600, 1200, and 1800 GMT in international Morse code. The messages are sent in an international surface analysis code that is entirely numerical and requires a set of tables for decoding. The appendix of the *Worldwide Marine Weather Broadcasts* contains an abridged decode for this numerical code; the complete weather code can be found in the *Federal Meteorological Handbook*, No. 2, "Synoptic Code."*

A representative of the National Weather Service in Washington suggests that the best way to receive these broadcasts is to listen to coastal stations for correspondence between them and a ship. Stations are listed in *Worldwide Marine Weather Broadcasts* with Class A1 for CW and Class A2 for a modulated emission.

A receiving station for the Cooperative Ship Program monitors many frequencies and the radio operator will shift frequencies until the receiving station has a good signal. Generally the broadcasts are sent over the high frequencies during the day and the lower frequencies at night.

Our friend William G. Konos tells us he listened to the U.S. Navy weather reports from station NAM in Norfolk, Va., in the Atlantic Ocean, and to station NPN in Guam, Marianas Islands, in the Pacific Ocean (among others) for their excellent weather reports during his years as a radio officer in the merchant marine and the U.S. Coast Guard. In addition to listening to broadcasts whenever Bill was at sea, he would make a "CQ" or "any ship" call for a ship about 200 nautical miles ahead of his vessel's course to ask about weather conditions.

---

* Purchase from: Superintendent of Documents, U.S. Government Printing Office, Washington, D.C. 20402.

CHAPTER 6

Anticipate Stormy Weather

Anticipate *sunny* weather for boating might be more to the point, since you have learned to make your own local weather forecasts based on as solid ground as scientific know-how and your own good weather senses permit. The reason behind weather study is to avoid having to deal with a bad storm when you are in your boat, if at all possible. But the "at all possible" is the crux of the matter; there are no weather seers around that we know of who have a 100 percent score in weather prediction.

Old hands who have experienced some fairly sticky weather conditions, at least according to the tales they tell, know the value of boating forethought in attitude as well as equipment. Several plans of action determined on and filed away for future use may mean the difference between having a whopper of a sea story to tell—or not.

ATTITUDE AND PREPAREDNESS

These two terms are inseparable as far as boating is concerned. There are certain pieces of safety equipment required aboard all boats by law, but it is impossible to legislate the right attitude. For example, we discuss later in this chapter several things you can do and try in case you find yourself out in a storm aboard your boat. Don't stop here, read all you can about these procedures in other books and articles. Talk to those who have tried them and ask for any tips they may have discovered that they think will be helpful to you. In other words, become familiar with safety procedures and then practice them aboard your boat with your accustomed crew time and time again in good weather until they have become routine.

By federal regulation, every boat is required to carry some type of flotation device, more familiarly known as a life preserver, for each person aboard. For some reason unknown to those experienced in the sport, novices to boating seem to feel wearing a life preserver shows them up as beginners and that they are to be worn only in case of dire emergency. Quite the contrary, although they are certainly to be worn in an emergency, signs of worsening weather make the wearing of life preservers a very acceptable idea. Furthermore, it is just ridiculous for anyone who cannot swim to go out in a boat without wearing a life preserver at all times except when they are below on a boat large enough to have a below. Small children should be trained from the start to accept this as part of the boating scene, like holding mother or dad's hand when crossing the street.

Responsible persons realize that no matter what size their boat may be, storms are potentially dangerous to the unprepared. It doesn't take long, as one sailing wife said, "to be up to one's bippy in water!"

It is important, therefore, to attend some of the many excellent boating courses given by the U.S. Coast Guard Auxiliary and the U.S. Power Squadron as well as to read the federal regulations on required equipment and *Navigation Rules* (formerly *Rules of the Road),** boating manuals, yachting magazines, fishing magazines, equipment catalogues, and the like. This continuing education process serves to keep you up to date on safety procedures and available equipment.

Confidence in yourself and in your ability to handle your boat is accepted as a matter-of-course attitude, but it is essential that this confidence be transmitted to everyone aboard—especially during a storm. Captain Blighs are still rampant on the seas. They may not be afraid themselves, but they transmit their tensions so effectively that everyone becomes nervous. In a bad situation, a sense of humor and a calm voice set everyone at ease and keep them functioning. As experience builds up, apprehension lessens.

We have two personal stories to illustrate this. Once when Sallie and Sam crewed on an overnight race, the night was pitch black and the winds were light. All the "laundry" on the racing yawl was flying: mainsail, mizzen, mizzen staysail, spinnaker, and spinnaker staysail. Though weather reports had mentioned only possibility of squally winds later, the skipper set the crew to removing and stowing the

* Obtain free of charge from an authorized chart agent or by writing to: Department of Transportation, U.S. Coast Guard (CG-169), Washington, D.C. 20590.

light sails when he saw lightning in the distance. A squall hit while they were trying to get the spinnaker down. The boat broached suddenly and water came in over the cockpit coaming and poured down the companionway into the cabin below. Sallie put her feet in the corner of the companionway in an effort to stem the flow, mentioning that the locker where the hatch slides were stowed was deep underwater. Sam calmly suggested that if she turned around and sat in the opening instead of putting her feet there it might be more effective, which she did and it was. All tasks were performed calmly and the emergency was soon over.

Aboard a small cruising boat, Virginia and two relatively inexperienced crew members found themselves in rough seas and high winds that came up without warning, despite the cheerful marine weather report for the area shortly before. Within minutes, they donned life preservers. As the rigging twanged shrilly in the steadily increasing wind, her friends became more and more apprehensive as they watched the spume whipping off the crests of the towering, breaking waves. The funny seems feeble now, but she calmed them by saying, "Don't be worried, we'll ride this out like a cork. The only real trouble we have is that my hand is so stiff from gripping this tiller, I may never play the violin on the concert stage again!" Privately, she confesses that as each wave seemed larger than the one before, she oft fervently repeated to herself the opening words of the Collect for Peace in the Episcopal Morning Prayer service, "Oh God, who art the author of peace and lover of concord."

It is also important to know when you are too tired to make sensible decisions and it is time for someone else to take over. Hallucinations are a problem if you stay up too long. One skipper who shall remain nameless sees little New England villages after a great many hours without sleep during overnight races. Since he knows he has reached his limit when this occurs, he mentions it immediately. His family always says, "Okay, that's it, time to get off the helm. We can now do better than you can."

Preparedness includes suitable clothing to be donned when weather conditions worsen. The following saga of the Pooles' first Christmas together illustrates this point, but also shows that when you try to explain some facet of your sport to landlubbers, you can find yourself up a creek without a paddle.

As they tell their story, they decided to give each other foul-weather gear and carted it on a plane to landlocked Ogden, Utah, to join Mary Lou's relatives for the holiday. Following this family's Christmas Day custom of opening one present at a time in turn before the assembled clan, Todd opened Mary Lou's gift to him first. His waterproof suit was a brilliant orange.

Mary Lou's mother exclaimed to her, "Dear, isn't that just a little vivid?"

"Oh, no," she responded, trying to placate her fashion-minded mother, "the bright color is so that Todd can be seen easily if he should fall overboard."

Next came Mary Lou's turn to open her new husband's gift to her. Her suit was a more becoming pale slate blue. A dreadful silence ensued!

Warm as well as waterproofed clothing is essential. The stronger the wind blows, the colder the air feels to the body despite the fact that the air temperature remains the same. This effect is called wind chill (see Appendix B).

SMALL BOATS IN STORMS

When you are out in a small boat and see a storm approaching too quickly for you to reach the safety of a harbor, everyone aboard should don a life preserver immediately. If the depth of the water is not too great, you can ride the storm out at anchor.

In deeper water, lower the sails on a small sailboat and try to steer the boat so that it runs before the wind under bare poles. Aboard a powerboat, reduce speed and head at an angle into the seas.

The weight of passengers and equipment should be kept low and distributed as near to the center of the boat as possible. In case of capsize, everyone should wear a life preserver and stay with the boat. In cold waters, move as little as possible, huddle yourself into a ball when alone or huddle together with others to conserve body heat. In particular, keep your outer clothing on because even though it is soaking wet, it still provides your body with some insulation.

MAN OVERBOARD

If at all possible, at least one other person should know how to handle the boat and how to retrieve someone who falls overboard. It is unpleasant to consider, but the person slipping over the side in rough weather just might be the only one with the experience and knowledge necessary to handle this type of emergency.

"Man overboard" is a serious cry and children should be firmly discouraged from using the term lightly. The throwable flotation device, such as a buoyant cushion or horseshoe life ring, required to be carried aboard any boat 5 meters (16 feet) or over, should be kept in an accessible place ready to be thrown at an instant's notice.

The man-overboard drill should be practiced by the whole family. We prefer to throw over a buoyant cushion and maneuver to retrieve it. Someone is stationed to keep an eye on it at all times, as should be done during an actual emergency.

There are a lot of true stories around to illustrate these points, and one of the most illustrative ones, we think, is the experience of a couple and their two young children in their small auxiliary under sail. The seas were very rough and the children were in their life jackets. However, the person who fell overboard was the un-life-jacketed father.

His wife did not even know how to start the engine, much less how to handle the boat under sail. Fortunately, which is the understatement of the year, she was able to attract the attention of persons aboard a passing boat. They immediately searched for and eventually rescued the father, who had managed to keep himself afloat in the meantime.

Another boat was flagged down and a member of its crew went aboard and helped her bring her craft safely into the nearest harbor. The family is still together because of an enormous amount of luck.

We find the following table relating water temperature "to approximate survival time of humans immersed in the sea" extremely interesting:

Water Temperature		*Exhaustion/ Unconsciousness*	*Expected time of Survival*
°C	*°F*		
0.3	32.5	15 minutes	15–45 minutes
0.3–4.4	32.5–40.0	15–30 minutes	30–90 minutes
4.4–10.0	40.0–50.0	30–60 minutes	1–3 hours
10.0–15.6	50.0–60.0	1–2 hours	1–6 hours
15.6–21.1	60.0–70.0	2–7 hours	2–40 hours
21.1–26.7	70.0–80.0	3–12 hours	3 hours–Indefinitely
26.7	80.0	Indefinitely	Indefinitely

This table appeared in a technical memorandum circulated by NOAA that further amplifies these figures by stating:

> The approximate survival time of human beings in the sea is directly related to sea surface temperatures . . . . The survival

> time given can be considered only a first order approximation since data records do not incorporate many uncontrollable physiological variables. For example, neither quantitative appraisal nor records have been made of such intangible factors as the physical condition of the individual or his will to survive . . . .
>
> Survival time also may be affected by body type, body attitude, and physical condition, amount of subcutaneous fat and will to survive. In waters warmer than 21°C (70°F), heat production may keep pace with heat loss, and fatigue leading to ultimate exhaustion is then the limiting factor.
>
> There are two principal schools of thought with regard to survival techniques during immersion: (1) vigorous exercise; and (2) passive waiting. Vigorous exercise may dissipate heat reserves more rapidly, though within reasonable time limits it may keep muscles warm and prevent their stiffening. Passive waiting maintains heat production by the natural process of shivering . . . . Passive waiting is somewhat in favor at present.

Since this article was written, extensive studies show that predicted survival times increase significantly when persons are relaxed and are in life jackets so that they expend as little energy as possible.

THE ROLE OF THE ANCHOR

In a list of safe boating tips, the U.S. Coast Guard states you should "have an anchor and sufficient line to assure a good hold in a blow." This is not the only way an anchor is useful in a storm.

When there is a possibility that the boat may be blown into an area of dangerous shoals, an anchor can be dropped over the side even though initially it just dangles because the line is not long enough to reach the bottom. Chances are excellent that as the boat is blown into shallower depths the anchor will catch and hold in time to save the boat from disaster.

Anchoring is often a wise choice when zero visibility threatens and your safety is at stake if you continue on your way.

NEAR ZERO VISIBILITY

Terror can strike your heart if you are isolated by near zero visibility and have not prepared for it beforehand. Sheets of driving rain can accompany a storm and blot out the world around you as completely as a heavy fog. This sort of curtain can surround a boat anywhere: in inland waters, along a coast, or at sea.

The skills for navigating in greatly reduced visibility should be readily at hand, having been practiced frequently in clear weather. When land is near there must be some method for measuring the depth of the water around the boat: an oar, a lead line, or an electronic depth sounder can warn you when you are off your course in confined waters. A skipper on any waters should have the ability to follow a compass course, know the compass has been corrected so that the readings are accurate, and have the proper chart or charts aboard. The position of the boat should be established before the storm hits. On seagoing boats so equipped, someone in addition to the helmsman should have the know-how to determine an accurate position by navigational equipment such as a radio direction finder, radar, loran, or the like.

Fog signals should be given from a boat as well as be identified when heard given by nearby boats that are not visible. Rules for these are presented in the *Navigation Rules* pamphlets, which are published in three editions: International-Inland, Great Lakes, and Western Rivers. For your convenience, we include a résumé of Inland Rules and note any difference when International or Great Lakes Rules apply:

Power Boats Under Way. 1 prolonged blast (4 to 6 seconds) at least every minute (international: 1 prolonged blast at least every 2 minutes. Great Lakes: 3 distinct blasts at least every minute).

Sailboats Under Way. Short blasts at least every minute: 1 blast signifies boat is on starboard tack, 2 blasts boat is on port tack, 3 blasts wind is abaft of the beam.

Boats at Anchor in a Fairway. A bell rung rapidly for about 5 seconds at least every minute. (Great Lakes: a bell rung rapidly for 3 to 5 seconds at least every 2 minutes; in addition, a signal of 1 short blast, 2 long blasts, and 1 short blast in quick succession at intervals of not more than 3 minutes).

OFFSHORE PREPARATION

Most of us dream of getting a larger boat and venturing further afield than we do now. Crowding masses of humanity at boat shows all over the country indicate the tremendous interest in cruising boats. The charter business is mushrooming in the United States and foreign waters. However, hand in hand with the long-range boating dream should be the stark realization that large bodies of water can generate very severe weather.

In severe weather conditions, as always, the primary consideration is the safety and well-being of the entire crew. The avoidance of "man overboard," injury, and seasickness (if possible) requires careful instructions, proper equipment, and other factors, including attitudes under very unpleasant conditions. Obviously the basic safety of the crew depends on the integrity of the boat; not only to avoid sinking or shipwreck but also to prevent damage and minimize the amount of water below decks.

The remainder of this chapter is directed at these considerations in enough depth to introduce the reader to this complicated subject. A well-designed, well-built boat, properly handled, can take more punishment than the crew. The noise and motion can be very disturbing; they should not be confused with the basic factors that are important to safety.

First there must be an honest evaluation of the seagoing capabilities of the structure and configuration of the boat as well as the equipment needed to withstand the rigors that might be in store. We do not intend here to discuss all of the kinds of equipment that should be considered, but those that we do may serve as eye-openers to what might be necessary under stormy weather conditions.

One of the concerns aboard a cruising boat underway is that persons may hurt themselves. A first-aid book and a well-stocked medicine kit should always be readily available. Included among the safety features considered essential to minimize the possibility of injury are handrails above and below deck and lifelines completely around the deck. In addition, safety belts as well as life preservers should be provided for each person aboard, to be worn especially for any work on the deck of a powerboat as well as a sailboat when the boat is pitching, plunging, or rolling in rough seas.

A fast self-draining cockpit is essential for any offshore cruising. Also, several methods for pumping the bilge should be available. The pumps should be in different locations. One should be accessible from the cockpit, one from below, and at least one adequate hand pump aboard in case of power loss. The logy condition of a boat that is full of water and wallowing contributes more to difficulty in boat handling than anything else.

Weather cloths are often included in the storm gear of seagoing boats as shields to keep as few waves as possible from coming into the cockpit. These pieces of heavy canvas or Dacron can be lashed to the lifelines and their stanchions so that when the weather cloths are in place, the height of effective freeboard is essentially raised to the height of the lifelines.

Procedures for coping with different types of leaks, such as those that might result from the hull working in a seaway, contact with sharp floating debris, or grounding in storm-lashed waters should be considered long before this sort of emergency occurs. Clothing can be used to stuff small leaks, spare boards or floorboards can be nailed over large holes, and boat mattresses can be wedged against the inside of the hull to help halt the onrush of water.

The best checklist we know for anyone who plans to cruise either along the coast or away from land appears in the equipment leaflet published by the United States Yacht Racing Union.* These lists are revised from time to time to include new equipment suggested by sailboat racing crews and race committee members to promote seaworthiness and safety. Although their purpose is to establish minimum equipment standard for sailing races, we feel they are of great value to persons who cruise or fish in power boats as well as sailboats.

Equipment recommendations in this list are divided into the following four categories:

1. Long distance offshore in open ocean where boat and equipment must be sufficient to withstand heavy storms and extended periods at sea
2. Along shorelines or in large bays or lakes with little protection and for extended periods of time
3. In open water which is relatively protected
4. Close to shore in relatively protected waters

SURVIVAL STORMS

A storm with winds of near gale or above and high breaking seas is particularly dangerous when it lasts for a long time, such as many hours or even days. Sometimes these conditions are referred to as "survival storms" and mariners find themselves in a lengthy struggle to do just that. They face the possibility of disabling fatigue suffered by everyone aboard when even the simplest task requires superhuman effort. Equipment aboard a boat is subjected to strains far greater than it was intended to withstand, and the likelihood of failure is great. Heading for a harbor may not be the safest thing to do when caught out in a storm of this type; a better alternative may be to head away from land and shoal areas.

Handling a boat to maintain sea room requires a careful balance between achieving the desired course and not driving the boat too

* Order from: USYRU, 1133 Avenue of the Americas, New York, N.Y. 10036.

hard. Where there is no navigational hazard other than the storm conditions to be considered, the person at the helm must judge when an increase or decrease in speed or a change in direction might make the difference between a dangerous situation and a merely uncomfortable one.

When the immensity of the waves and the strength of the wind make it impossible to continue in a normal fashion under power or sail, some other method of coping with these conditions must be adopted. One of the best ways to gain knowledge of the techniques of seamanship that can be tried during a survival storm is by reading articles and books about and by those who have experienced and survived them. For example, *Heavy Weather Sailing** by K. Adlard Coles is an excellent collection of reports by racing sailboat skippers describing methods used to withstand different storms. Useful information is also found in the "Handling at Sea" chapter in *Naval Shiphandling*† by Captain R. S. Crenshaw, Jr., USN, a book used for training U.S. Navy personnel. We also recommend several discussions in *Voyaging Under Sail*‡ by Eric C. Hiscock that pertain to this subject.

You will discover in your reading that several different methods of boat handling are described. In the accounts, each method used has probably worked successfully because it was suited to the hull design of the boat used in the waves prevailing at that time, and it might meet with less success if used with another design under similar conditions. Since many of the boats featured in these accounts may differ from yours, trying out various methods in your own boat during windy weather should give you some idea of how your boat is likely to react in heavy seas and help you to determine the methods most suitable for a design like yours. A skipper who is aware ahead of time of the problems he may someday face is likely to make more effective on-the-spot decisions.

Most of the suggestions for boat handling under extreme weather conditions fall into one of four general classifications. These are lying a-hull, modified lying a-hull, heaving-to, and running before the wind.

LYING A-HULL

Some experienced boatmen with seagoing boats advise doing nothing to resist the force of the seas as conditions worsen. They stop

---

* Published by John De Graff, Inc., Tuckahoe, N.Y. (reprinted 1969).

† Published by the United States Naval Institute, Annapolis, Md.

‡ Published by the Oxford University Press, London and New York (reprinted 1963).

the engine, lash the helm, and drift. On a sailboat, all sails are lowered.

Once everything on deck is secured or stowed, everyone aboard goes below. The ports and hatches are firmly latched. Even though conditions in the closed-up cabin may be rather grim, everyone has an opportunity to get some essential rest. This is a big advantage of lying a-hull.

The boat is likely to assume a position that is more or less broadside to the wind and waves. Sometimes, as the boat drifts, the water through which it passes is smoothed slightly, which seems to exert a calming influence on the breaking waves immediately to windward.

Experience has shown that lying a-hull can be dangerous for relatively small boats not designed for seagoing use. These have been known to be capsized by exceptionally large breaking waves. Power boats with shallow drafts relative to their high superstructures may be particularly vulnerable to capsize. On the other hand, deep-draft boats may ride out a storm more easily using this method. The experience of the captain of a freighter in the Military Sea Transportation Service during World War II seems to bear this out. He reported that when he lost all power aboard his freighter during a hurricane, he noticed to his surprise that the ship rode the seas with less strain than when the engines were running and he could keep the bow into the waves. The next time he was caught out in a hurricane, he deliberately cut power and had the same experience.

If there is any possibility that the wind will blow the boat on shore before the storm is over, lying a-hull should not be considered. The helmsman should remain in control.

MODIFIED LYING A-HULL

As noted before, the stability or construction of some boats makes it unsafe for them to lie abeam to the seas. Although the U.S. Navy considers a 30-degree roll an acceptable amount to be expected when a boat is at sea, this might be dangerous for some recreational boats that could be rolled over enough to expose their many-windowed deckhouses to the full force of the breaking seas. With this kind of hull design, something needs to be done so that the boat is kept either bow-to or stern-to the waves.

Since a boat is designed to ride bow into the seas it seems that the wisest decision would be to have this part of the boat exposed to the fury of the waves first. Unfortunately, few if any boats will ride with their bows toward the wind and waves unless you do something to help them, and even then you have an expletive deleted time keeping any bow in this position.

It is often possible to hold the bow of a boat more toward the wind and waves by letting out some type of drag from it, such as long heavy lines or, as advocated by some, a sea anchor attached to the end of one long heavy line. The most common sea anchor is a canvas cone with a wire loop or two crossed pieces of wood to hold the large end open. (Mattresses, floorboards, and the like can be substituted for a commercially marketed sea anchor.) The heavy line attached to the sea anchor should be secured to a strong fitting, which in turn should be bolted through the deck of the boat as there will be a tremendous strain on the fitting, the line, and the sea anchor itself.

Opinions concerning the value of commercial sea anchors for cruising boats are mixed, but the majority vote seems to be against their use—mainly because of the large sizes required if they are to be really effective. There have been cases when boats, held with their bow toward the waves by inadequate-sized sea anchors, have been pushed backward by the force of the wind so quickly that their rudders have been severely damaged. The value of sea anchors when used aboard lifeboats, life rafts, and dinghies, however, has been amply demonstrated.

The effectiveness of keeping the bow head-to-wind appears to be a function of the draft of the boat. An extremely large sea anchor is usually required to hold the bow of a deep heavy boat toward the direction of the wind. Sometimes a small steadying sail in the stern, such as is used on yawls, ketches, or seagoing fishing boats, helps to maintain a boat in this position.

Many boats will ride better in heavy seas when the drag is off the stern. The strength of the transom as well as that of the cleats on it must be taken into account, however, for this section of the boat is rarely designed and constructed to withstand this type of treatment. A single long heavy line is a particularly effective drag when the ends are secured to each side of the transom so that it is streamed astern in an elongated loop.

In some weather conditions it is possible for everyone to be below when a stern drag is used, if the tiller or wheel can be lashed in such a way as to steer the boat so that the waves will be taken on the stern quarter. The decision as to which way to try to head the boat sometimes depends on the imminence of land, shoals, or other similar hazards in addition to sea conditions.

Having a boat ride stern-to the waves has its dangers too, as there is the possibility that a wave may break over the stern. When this happens, the boat is said to be pooped and may be in peril. A breaking wave can fill the cockpit and even smash hatch covers, doors, or windows, making it possible for green water to pour into the hull. Until most of this water is pumped out, the stability of the boat is seriously impaired, causing it to wallow sluggishly in the seas.

HEAVING-TO

Under some stormy conditions, it is possible to maintain enough steerageway under power toward the waves without making appreciable forward progress through the water. The purpose of this maneuver is so that the boat offers the least possible amount of resistance to the seas.

The success of this method depends on the skill of the helmsman who must constantly decide whether to increase or decrease speed or to turn away from or toward the waves as the situation at the moment requires. Aboard a powerboat or an auxiliary, the length of time this method can be used may depend on the fuel supply. As time wears on, it may be necessary to adopt some other method of coping with these conditions in order to conserve fuel because of the possibility of a later, greater emergency.

Aboard sailboats, heaving-to can be accomplished with a small amount of sail. (It can be practiced in good weather by using more sail area.) A small sturdy jib is sheeted slightly to windward so that it is actually pulled over on what would be the wrong side of the boat under normal sailing conditions. At the same time, a reefed mainsail or mizzen is sheeted in and the helm is lashed down to steer the bow toward the wind. The wind blowing on the sails will cause some forward motion, but as it blows on the backed jib the bow is pushed to leeward or away from the wind. Then, the positions of the aft sail and the rudder counter this by tending to turn the bow toward the wind once more. Thus, if everything is balanced correctly according to the forces present, the boat should head up toward the wind, fall off away from the wind, head up, fall off, and so on while making a slight forward progress through the waves. The boat should take care of itself except for occasional adjustments as sea conditions change.

When the waves become so high that the boat is cut off from the wind when it is in the troughs of the waves, heaving-to under sail will have to be abandoned. The sails slatting in the troughs and then suddenly filled by the full force of the wind as the boat is lifted to the wave crests can become dangerous instead of helpful.

RUNNING BEFORE THE WIND

As the storm worsens, the boat is often steered so that it is said to run before the wind without power on a powerboat or under bare poles on a sailboat. Boats will usually suffer less strain and conditions will be more comfortable for the crew when the stern quarter or stern is presented to the seas.

Once again, the success of this method is a measure of the skill of the helmsman. If the boat is pushed along by a towering, onrushing wave, the stern is lifted and the boat hangs for an instant on the crest. It will race uncontrolled down the steep watery slope if the helmsman is not careful. By reacting with precise timing to the motion of the boat, an experienced helmsman can turn his craft slightly to one side or the other, slowing its downhill trip by quartering the wave with a slight sideslip somewhat like a good skier traverses as he descends a steep, wide mountain slope.

There is a possibility that under some weather conditions the boat will pick up too much speed despite all the efforts of an alert helmsman. The boat is now in danger of being pitch-poled by a wave. This happens when a boat surfs down a wave so fast that it buries its bow in the trough and is picked up by the stern and flipped over, end for end. When this becomes a possibility, steps must be taken at once to slow the speed of the boat. Long heavy lines dragged from the stern should retard this headlong progress. These lines should be assembled earlier in readiness for just such a probability.

Boatmen have found that they have been able to exercise some control over the speed of their boats by varying from time to time the number and lengths of dragged lines as wind and wave conditions change. Aboard small boats, where the number of long heavy lines may be limited, floorboards, cockpit cushions, mattresses, and the like can be tied to available lines and streamed overboard.

As with all the methods discussed, whether or not to drag lines from the stern requires an on-the-spot decision according to the particular combination of weather conditions at the time. A drag such as this has been said by some to retard the lift of the stern and thus increase the danger of being pooped as described under modified lying a-hull. On the other hand, others say that setting a drag tends to steady a boat, making it easier for the helmsman to keep the stern toward the seas, which lessens the danger of broaching, ever-present when running before large waves.

Broaching can occur when a wave that is sometimes out of step with its brothers throws the stern to one side so that the boat is suddenly broadside to the wind and sea and is rolled dangerously, possibly far enough to overturn. Awesome as this is to visualize, you may find some comfort in the fact that many sailboats designed for ocean passages have righted themselves after a capsize. Their bedraggled owners have reached shore singing the praises of the designers and builders of these boats.

COMPETENCY IN HANDLING A BOAT

It is impossible to make a judgement ahead of time as to exactly what steps to follow when caught out in conditions of extreme weather. So much depends on the force of the wind, the size and frequency of the breaking waves, the design and idiosyncrasies of a particular boat, and the experience of the crew. The main thing that nearly everyone who has ever faced nature at her nastiest at sea advises is continual reading about the experiences of others and the techniques they have used. Then, if you are caught out in a survival situation and one method does not work, you will at least know of others to try. Those who have lived to tell the tale usually agree that trial-and-error experimenting is the only way to proceed.

The possibility of coping with a survival storm, however remote it may now seem, should provide all boating enthusiasts with sufficient incentive to learn as much as they can. Captain R. S. Crenshaw, Jr., USN, says in his book *Naval Shiphandling,* "The best preparation for handling the ship, then, is a combination of a study of the principles involved and sufficient experience at sea to be able to evaluate the situation. Neither extensive experience nor theoretical understanding can stand alone. The competent shiphandler must have both."*

CHAPTER 7

Probe into Nature's Whirlwinds

Woven into the fabric of legends of the sea are tales of nature's awesome windstorms—the mighty tropical hurricane and the sea's sister to land's destructive tornado, the waterspout. When threatened by these furies, a mariner becomes all too aware of his insignificance.

Each year, however, these mysterious storms yield more of their secrets to meteorologists who pursue the study of them with relentless determination. But no matter how much knowledge is amassed, man and his creations must still try to avoid them or bend before them.

HURRICANES

As more information became available about weather patterns around the globe, meteorologists began to understand some of the conditions that cause the lethal and terrifying storms called hurricanes. The advent of weather satellites that orbit the earth has enlarged their scope of prediction and study. They carefully study cloud photographs taken from the satellites to detect whether any tropical storms may have formed that might develop into hurricanes.

From the vantage point of a satellite, a tropical cyclone north of the equator appears as a spiral cloud formation that rotates in a counterclockwise direction. Hurricanes that most frequently affect the United States and its coastal waters originate in the warm waters of the western Caribbean Sea, the Gulf of Mexico, and between latitudes 5°N and 20°N in the tropical North Atlantic Ocean. Hurricanes originating in the eastern North Pacific Ocean between latitudes 10°N and 20°N westward from the North American coast can hit either southern California or occasionally even the Hawaiian Islands. The islands are less likely to be struck by typhoons, the term used for hurricanes spawned in the tropical waters of the North Pacific Ocean west of longitude 180°.

Hurricanes are not associated with any frontal systems. They are steered instead by the wind currents around the large high pressure systems that dominate the tropical and subtropical oceans. When the wind in a tropical cyclone reaches a speed of 64 knots or more the storm is officially called a hurricane.

Hurricane Advisories As a hurricane develops, hurricane advisories for the Atlantic, Caribbean, and Gulf of Mexico warning system are issued by the National Hurricane Center in Miami, Fla. Shipping in the Pacific receives warnings from centers in San Francisco, Calif. and Honolulu, Hawaii. The National Weather Service informs the public on radio and television broadcasts as to the intensity, location, and probable track of the hurricane.

According to Dr. Neil Frank, director of the National Hurricane Center:

> We can give several hours notice (12–24 hours) of an approaching hurricane . . . . We can also predict the future track of a hurricane with reasonable accuracy. Our normal forecast error for 24 hours is about 100 miles. This means marine interests do get enough advance notice with a hurricane to take action.

The informative pamphlet *Hurricane** describes the special hurricane warnings that are issued to the public as follows:

> *Small-craft advisory:* When a hurricane moves within a few hundred nautical miles of the coast, advisories warn small-craft operators to take precautions and not to venture into the open ocean.
>
> *Gale warnings:* When winds of 33 to 48 knots are expected, a gale warning is added to the advisory message.
>
> *Storm warning:* When winds of 48 to 64 knots are expected, a storm warning is added to the advisory message.
>
> Gale and storm warnings indicate the coastal area to be affected by the warning, the time during which the warning will apply, and the expected intensity of the disturbance. When gale or storm warnings are part of a tropical cyclone advisory, they may

---

* Order from U.S. Department of Commerce, NOAA, National Ocean Survey, Rockville, Md. 20852.

change to a hurricane warning if the storm continues along the coast.

Hurricane watch: . . . A hurricane watch means that hurricane conditions are a real possibility; it does *not* mean that they are imminent. When a hurricane watch is issued, everyone in the area covered by the watch should listen for further advisories and be prepared to act quickly if hurricane warnings are issued.

Hurricane warning: When hurricane conditions are expected within 24 hours, a hurricane warning is added to the advisory. Hurricane warnings identify coastal areas where winds of at least 64 knots are expected to occur. A warning may also describe coastal areas where dangerously high water or exceptionally high waves are forecast, even though winds may be less than hurricane force.

When a hurricane warning is issued, all precautions should be taken immediately. Hurricane warnings are seldom issued more than 24 hours in advance. If the hurricane's path is unusual or erratic, the warnings may be issued only a few hours before the beginning of hurricane conditions. Precautionary actions should begin as soon as a hurricane warning is announced.

The Hurricane Season Along the Atlantic and Gulf of Mexico coasts, the season for these storms corresponds generally with the time of the greatest boating activity. Even though the larger percentage of hurricanes occurs in September, with August and October vying for second place, these storms have occurred as early in the year as May and as late as November.

The hurricane season is officially defined as follows:

Atlantic, Caribbean, the Gulf of Mexico, from June through November

Eastern Pacific, June through November 15

Central Pacific, June through October

Paths of Hurricanes It is difficult to make accurate predictions concerning the paths of hurricanes because no two hurricanes ever follow exactly the same track.

While a hurricane is in tropical waters, it is influenced by the Northeast Trade Winds and moves toward the west or west-northwest at a speed of approximately 10 to 15 knots. Near latitude 30°N, the Prevailing Westerlies usually take over and cause the storm to change its direction and curve toward the northeast. (On land, lat. 30°N cuts across the northern part of the Florida peninsula and across Louisiana south of New Orleans.)

After the recurvature (when the storm curves northeastward), the speed of a hurricane is likely to accelerate to 15 to 30 knots and sometimes as high as 50 knots. It continues on its path until it either reaches cold waters or land, when it diminishes rapidly. The large Bermuda-Azores high pressure area, which is ordinarily near Bermuda during the summer and autumn months, can move northward or northwestward from its normal position where it deflects the usual path of a hurricane and sends it toward the eastern seaboard of the United States.

Although an understanding of these probabilities may help you decide the best course of action to follow if you are so unfortunate as to be caught out in a hurricane, there are no sure bets when hurricanes are concerned; some are very erratic and never recurve, and others drift aimlessly around, even forming complete loops before they die out.

Visual Hurricane Signs As with all low pressure areas north of the equator, the air within a hurricane whirls counterclockwise in toward the center of the low pressure area, moving even faster as it approaches the center. (This is the spiral cloud formation that can be seen in satellite photographs of tropical cyclones and fully developed hurricanes.) Before reaching the comparatively calm central eye or core, the air is thrown upward and windswept cirrus clouds are thrust out from the center of the storm, usually in a different direction from the surface winds. These cirrus clouds can be seen over an area with a diameter as large as 350 to 450 nautical miles and therefore can be the first visual signs available to a boatman.

Ocean swells also indicate that a hurricane might be imminent. These radiate from and precede an approaching storm. The larger and better-developed the storm, the farther the swells will outrun it. Swells will be most evident to the right of the center of the storm relative to the direction in which the storm is moving. Here, the speed of the hurricane wind is added to the terrestrial winds that drive the storm along its course.

As the hurricane nears your location, there is a dramatic drop in barometric air pressure (record low readings of below 92 kiloPascals (27 inches of mercury) have been recorded in the centers of some especially severe storms). Strong winds can be expected over an area of 80 to 180 nautical miles in diameter. These winds increase to most violent proportions in an approximately 9- to 10-nautical-mile-wide ring surrounding the eye of the storm, and the wind speed here can be from 65 to 180 knots.

The spiral cloud culminates in this ring, which is referred to as the eye wall. Torrential rains are released by the clouds, but in the areas between them there may be only light rain or none at all.

Inside the eye of the hurricane, however, the wind speed may be only 15 knots or less. The eye of the hurricane is generally 12 to 22 nautical miles in diameter. Here, blue sky and bright sunlight can often be seen overhead through a layer of thin or broken clouds. The seas within the eye are high and confused.

Encountering a Hurricane Mariners unfortunate enough to have been caught in the path of a hurricane report that it is infinitely easier to handle a boat in the ever-increasing winds as the eye approaches than in the sudden onslaught of winds from the opposite

Fig. 12. GENERAL PATHS OF HURRICANES—Gulf and Atlantic Coasts *Shaded area in each hurricane representation indicates the right or dangerous semicircle and the arrowheads its counterclockwise rotation.*

The long arrow shafts show the general paths of hurricanes.

The probable speed of travel of these storms is 10 to 15 knots south of latitude 30°N; after recurvature, the speed of travel is usually 15 to 30 knots (can be up to 50 knots).

direction when the eye has passed. It is important for boatmen to avoid the eye and, as much as possible, to steer away from the path of the hurricane.

As the speeds of the winds in the *right* half of the storm (right relative to the direction in which the storm is moving) are added to the speeds of the terrestrial winds propelling the storm, it is vital to avoid the "right or dangerous semicircle." In the *left* half of the storm, the terrestrial winds responsible for the travel speed of the hurricane counteract the speed of the hurricane winds, and thus it is better to be in the "left or navigable semicircle" where the winds and seas are comparatively less violent.

Because the winds spiral in a counterclockwise direction in a hurricane north of the equator, Buys-Ballot's Law can be applied to locate the center of the storm. When you face into the wind, the center is to your right and slightly behind you. If you are unable to get a report of the direction in which the storm is traveling, you will have to assume it is traveling in a generally northwesterly direction when it is south of latitude 30°N and in a northeasterly direction after a recurvature north of this latitude.

Although making forward progress within a hurricane may be difficult and often impossible, we offer the following general rules for power-driven vessels to avoid a hurricane storm center in the Northern Hemisphere, as summarized in NOAA's publication *U.S. Coast Pilot:*

> *Right or dangerous semicircle:* Bring the wind on the starboard bow (045° relative), hold course and make as much way as possible. If obliged to heave-to, do so with head to the sea.
>
> *Left or navigable semicircle:* Bring the wind on the starboard quarter (135° relative), hold course and make as much way as possible. If obliged to heave-to, do so with stern to the sea.
>
> *On storm track, ahead of center:* Bring the wind two points on the starboard quarter (157½° relative), hold course and make as much way as possible. When well within the navigable semicircle, maneuver as indicated above.
>
> *On storm track, behind center:* Avoid the center by the best practicable course, keeping in mind the tendency of tropical cyclones to curve northward and eastward.

Our interpretation of these rules for maneuvering in a hurricane are illustrated in Fig. 13.

Following are suggestions, also from the *U.S. Coast Pilot,* for methods to be used by powerboats and sailboats when it becomes necessary to heave-to (to maintain steerageway without making ap-

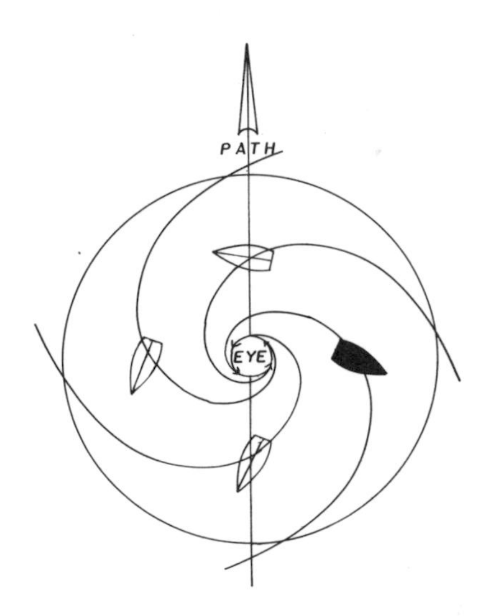

Fig. 13. MANEUVERING IN A HURRICANE—Our interpretation of the general rules as presented in the *U.S. Coast Pilot* *The suggested heading is shown for boats in different relationship to the eye of the storm. They are drawn on wind arrows spiraling counterclockwise toward the center of the low pressure area. The shaded boat is in the right or dangerous semicircle.*

This drawing may be rotated so that the line depicting the path of the hurricane is lined up in the same direction as the path of an existing hurricane. By doing this you can determine from the direction of the wind at your location your probable position within the storm.

preciable forward progress through the water) when they are within a tropical cyclone.

> If it becomes necessary for a vessel to heave-to, the characteristics of the vessel should be considered. A power vessel should be concerned primarily with damage by direct action of the sea. A good general rule is to heave-to with head to the sea in the dangerous semicircle or stern to the sea in the navigable semicircle. This will result in the greatest amount of headway away from the storm center, and least amount of leeway toward it. If a vessel handles better with the sea astern or on the quarter, it may be placed in this position in the navigable semicircle or in the rear half of the dangerous semicircle, but never in the forward half of the dangerous semicircle. It has been reported that when the wind reaches hurricane speed and the seas become confused, some ships ride

> out the storm best if the engines are stopped, and the vessel is permitted to seek its own position. In this way, it is said, the ship rides with the storm instead of fighting it.
>
> In a sailing vessel, while attempting to avoid a storm center, one should steer courses as near as possible to those prescribed above for power vessels. However, if it becomes necessary for such a vessel to heave-to, the wind is of greater concern than the sea. A good general rule always is to heave-to on whichever tack permits the shifting wind to draw aft. In the Northern Hemisphere, this is the starboard tack in the dangerous semicircle and the port tack in the navigable semicircle.

If you are in a harbor when a hurricane is on the way, attention must be given to the mooring of your boat. Considerable destruction is caused along a coast by a very high rise in the level of the water. This can reach as much as 20 feet above normal. High waves riding on top of the storm surge add to the potential for destruction.

Fortunately, the U.S. Corps of Engineers has designated many existing protected areas along the seacoasts and in rivers and creeks as hurricane holes; in some cases these areas have been dredged and further safeguarded by the construction of jetties and the like. Persons who live near the water in places where hurricane holes have not been designated can often tell you which rivers, inlets, coves, and creeks will provide the best protection for your boat.

Additional lines should be put out and extra scope should be added to all mooring lines. Lines should be as strong as possible and at least one should be of chain. Chafing gear should be used around all rope lines wherever they make contact with chocks or any other part of the boat.

A fisherman in the Bahamas told us that he always takes his boat to one particular creek when a hurricane is forecast and puts out three anchors around his boat. He then swims down to the anchors to be sure that one fluke of each is hooked under coral. He checks to make sure that at least one anchor will hold no matter which way the wind blows. Although coral is not available everywhere in hurricane waters, his three-anchor-and-fluke-hooking idea is worth considering, especially where the bottom is rocky.

According to the National Weather Service, tornadoes spawned by hurricanes are as destructive as the storm winds and seas. Therefore, persons anywhere in the vicinity of the probable path of a hurricane should keep listening to the radio for a possible announcement of a tornado watch or warning.

WATERSPOUTS

The violent, contained, and capricious windstorms called tornadoes are dreaded by inhabitants over wide areas of the earth's land surface. Their less well-known sea sisters, waterspouts, have the capacity for being just as violent; however, they tend to move more slowly and last a shorter time. The destruction caused by these storms has been comparatively small. Because of the uncrowded nature of the sea where waterspouts range, they rarely engulf mariners and their vessels. Thus, their fury is usually vented by harmlessly churning up great volumes of water.

A waterspout is a circular fountain of spray caused by a spinning elongated column of air descending from a cloud overhead. Potential waterspout activity starts as a swirl on the underside of a turbulent cumulus cloud, and, as it develops, a funnel-like form grows down from the base of the parent cloud. Although many potential waterspouts dissipate before they are fully formed, those that reach the mature stage lengthen until the parent cloud is joined by the funnel to the waters below.

The air in the outer layers of a mature funnel cloud spirals rapidly upward around a wall of condensation. This wall does not seem to move, whereas inside it there is a core of low pressure where the air descends slowly. Condensation is caused by the spiraling rising air that cools and condenses around the wall into a cloudlike form visible to the human eye.

The surface of the water beneath the parent cloud starts to rotate long before you can sight the meeting of the funnel cloud with the water. The forces in the column of air of a developing waterspout increasingly raise the water in a circular pattern until it contacts the lengthened funnel cloud. The visible spray whirling into the air is the actual waterspout. Although the term waterspout is considered generally to encompass the entire series of events, some scientists refer to the funnel cloud and the water circulation as separate parts of a single funnel event.

Most meteorologists agree that the general identification of waterspouts refers to two types. Tornadic waterspouts are spawned by the same kind of weather that generates tornado activity. They can form either during the turbulent conditions of prefrontal squalls that are likely to precede a rapidly moving cold front or when a strong cold front collides with a moist warm front.

Waterspouts of the second type, or common waterspouts, are observed more frequently. They are not formed as the result of a frontal system and sometimes are sighted in fair weather.

Although common waterspouts have been sighted off all the coasts of the United States and on the Great Lakes, they are most likely to occur in southern waters during the warm, humid, rainy summer season. Intense heating of the water takes place during this season, especially the water in shoal areas around islands and in shallow bays.

Observers at a safe distance from this type of waterspout said they were in winds that tended to be gentle, usually less than 10 knots. Although the prospects for rain were generally high at the time of many reported sightings, records show thunder and lightning were evident during only about 30 percent of the events.

Records of vessels where waterspouts of either type were actually encountered indicate that waterspout activity produces high wind usually accompanied by heavy rain and an abrupt drop for a few seconds in barometric air pressure while the eye of the waterspout is directly overhead.

Visual Signs Aloft Dr. Neil Frank, director of the National Hurricane Center, points out that "there is a big difference in the time lead for warnings for hurricanes and waterspouts . . . the lead time for waterspouts is seldom more than a few minutes." Mariners must therefore rely on visual signs aloft to warn of possible waterspout activity.

A line of heavy, dark-based cumulus clouds are almost always overhead during periods of waterspout activity. As these clouds may be caused either by cold-front activity or local warm, humid conditions, they are likely to be present when either type of waterspout is sighted. We have observed that waterspout activity is a possibility when the lower edge of a cumulus cloud bank is ragged looking with dark curling clouds hanging from it, some straggling down further than others. Although the majority of these wisps never elongate enough to touch the water, each one is a potential waterspout. A cloud wisp can develop before your eyes into a smooth-sided funnel resembling most closely an elephant's trunk and moving in the sky as it lengthens with the same sweeping, questing motion. Under the cloud base, an eerie many-toned bluish-gray light, which sometimes has an overlay of yellow or green, often makes the outlines of the curling wisps and lengthening funnels more pronounced.

Reports by mariners of multiple waterspout activity are often recorded by the National Weather Service. Our frontispiece is a NOAA photograph taken in the shoal waters of the Bahama Islands in May 1961 that dramatically illustrates multiple waterspouts in action.

Until recently, there has been very little detailed information available on waterspouts. We are very grateful, therefore, to Mr. A. I. Cooperman, chief of NOAA's Data Information Group for collecting and sending us several articles on the subject. We are particularly indebted to Dr. Joseph Golden, research meteorologist at NOAA's National Severe Storm Laboratory for sharing with us so many of his latest findings on waterspout behavior.

Stages of the Common Waterspout In a scientific release by NOAA, Dr. Golden identifies five stages in the life cycle of a fully developed waterspout. This research is primarily concerned with common waterspouts observed off the Florida Keys. According to Dr. Golden, common waterspouts occur more frequently in this area than anywhere else in the world.

Dr. Golden describes the five waterspout stages as follows, with the quotations taken directly from a NOAA release.

The *Dark Spot* in the water is the first stage. This is not always accompanied by a funnel cloud although often a funnel is evident under the formation of cumulus clouds. Air above the dark spot "has a rotating circulation with a maximum speed of 20 to 30 knots in a broad circular band some 70 feet* from the center of the spot . . . . The observed duration of this first stage ranged from 1 to 22 minutes."

The *spiral pattern* of the second stage has a spiral of light and dark water around the dark spot. In this stage the waterspout intensifies and enlarges. "The dark spot is of the order of 50 to 200 feet* in diameter while the spiral pattern ranges from 500 to 3,000 feet† in diameter on the sea surface . . . . The duration of this stage was observed to range from 2 to 7 minutes."

The *spray ring*, or third stage, picks spray off the surface of the water indicating winds above 45 knots. The "funnel cloud often doubles in diameter and lengthens down to the sea surface . . . . The base of the waterspout 'marches' away from the slower parent cloud overhead, usually taking a course away from a neighboring shower . . . . The spray-ring stage is the shortest, lasting only a minute or two."

The *mature waterspout*, the fourth stage, results in the greatest diameter and length of the funnel cloud, "linking the sea surface and the parent cloud . . . . Rotational speeds in the belt of maximum winds around the eye of the spray vortex may be as high as 170 knots

---

* 21 meters.
* 15 to 60 meters.
† 150 to 900 meters.

. . . . On the surface, the spiral pattern shrinks, and forward motion of the waterspout reaches a typical maximum of about 5 to 15 knots, and occasionally up to 25 to 30 knots for short periods of time . . . . The observed duration of the mature stage ranged from 2 to 17 minutes."

The *decay stage* is the fifth and final stage. It "generally occurs abruptly as an advancing rainshower begins to overtake the waterspout from the rear, the . . . funnel shortens, tapers, and often assumes odd twisting shapes before it vanishes altogether."

Dr. Golden's findings included the observation that the average duration of funnel events (including both funnel clouds and waterspouts) was 14.6 minutes. Multiple funnel sightings with two or more occurring simultaneously, or more than one event in the same day, occurred in about one-third of the days when funnels were sighted in the Keys.

Avoiding a Common Waterspout Once the path of a common waterspout is observed, it is usually possible to avoid it by heading as rapidly as possible on a right-angle course from it. When cruising in warm waters, it is important to keep a weather eye peeled for cloud formations and water disturbances so you will not be caught before it is too late to take evasive action.

Waterspout conditions are not easy to forecast. Dr. Golden says, "Waterspout weather in the Keys to most laymen would appear to be normal, undisturbed summertime tropical weather with light surface winds, high surface temperatures and humidity and long lines of building cumulus congestus clouds with flat, dark bases."

As this type of weather is also prevalent during the summer in the southern states on the Atlantic coast and the Gulf of Mexico, these waters offer likely breeding grounds for the common waterspout. Although the common waterspout can form in all boating areas of the United States, Dr. Golden observes that it occurs least often along the Pacific coast and in the waters off Hawaii.

He also advises that whenever a boat is in an area threatened with waterspout activity, all loose items should be secured. One of the greatest hazards for boaters from the waterspout spray vortex seems to be flying debris from the boat itself. When dealing with the common waterspout, he adds, "Common sense is by far the best guideline."

Avoiding a Tornadic Waterspout Waterspouts of tornadic origin are one more serious boating hazard during the severe storm conditions that accompany them. This type of waterspout may also occur in any of the boating areas in the United States including Hawaii, but are observed most frequently in the Great Lakes and off the coast of the Gulf of Mexico.

Like tornadoes, these waterspouts are spawned by prefrontal or frontal squalls and, therefore, studies carried on by the National Weather Service concerning tornado behavior apply. These studies reveal that tornadoes can occur any time of the year, but they are more common in the United States during the customarily unsettled weather of the spring season. Following are excerpts from National Weather Service reports for the three areas where boatmen are most likely to be affected.

Central Gulf States: Tornado frequency begins to increase in February and to subside after May.

Southeast Atlantic States: Tornado activity increases in March and reaches its peak in April.

The Great Lakes: Most tornadoes occur in this area in June.

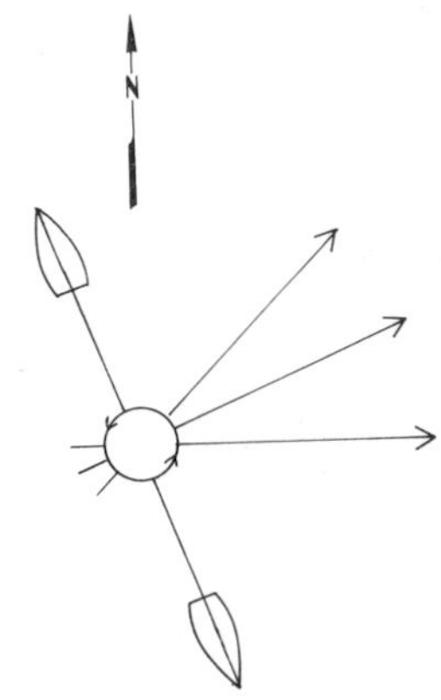

Fig. 14. AVOIDING A TORNADIC WATERSPOUT *The suggested courses are based on right-angle routes from the probable east to northeast travel path of a tornadic waterspout.*

The average speed of these storms is 25 to 30 knots (can be as much as 60 knots).

"Waterspouts also occur," Dr. Golden states, "over large inland lakes and off the eastern U.S. coast during late fall and winter during major cold outbreaks . . . . The waterspouts of tornadic origin most often move towards the east to northeast, but may move toward most any direction, north-east-south. They rarely move at constant forward speed, even while over land, and average forward speeds of 30–40 mph* and as high as 60 mph† for short periods of time."

The National Weather Service keeps a very close 24-hour watch when atmospheric conditions are right for the formation of funnel clouds, and radio bulletins are issued whenever waterspouts are sighted. When tornadic conditions are expected to develop, a tornado watch is announced over radio and television stations. A tornado warning is given when a tornado is sighted on a radarscope or a report has been received of a visual sighting. All personnel involved in tracing the storm or in providing protection and aid for persons in the danger area are notified. For example, a representative of the National Weather Service Forecasting Office in Cleveland, Ohio, told us, "Tornado watches and warnings are relayed to the Coast Guard when any part of the Great Lakes is affected. Also such messages are included in VHF-FM radio broadcasts."

In a *Fifth District Coast Guard Bulletin*, the encounter of U.S. Coast Guard cutter *Chilula* with a mature waterspout is described. The incident occurred during the month of February in the Atlantic Ocean about 170 nautical miles east of Norfolk, Virginia. On that day, the National Weather Service noted a cold front almost over Norfolk.

Those aboard the *Chilula* estimated the diameter of the waterspout to be 45 meters (150 feet) at the base. As the spout passed over the ship, a wind speed of 80 knots was measured. One of the crew recalled that as the center passed over it was "like looking up a chimney." Fortunately, this awesome whirlwind caused very little damage to the cutter.

Closer to home, and too close for comfort, Virginia's husband and his brother each witnessed tornadic waterspout activity from the same frontal system. At about 4:30 P.M. on a July afternoon, John Ericson was sailing his small, day-sailing Bullseye to its mooring in the protected harbor of Salem, Massachusetts. His brother Bill was about four nautical miles to the east and about one nautical mile off Gloucester Harbor heading for Marblehead Harbor aboard his auxiliary sloop.

John was well into the harbor trying to outrun a storm approaching from the west. He reports:

> I saw lead-gray streamers tapering down from a rapidly building cloud formation. The sky was very dark and emitted a yellow-green diffused light which lit up the white boat hulls and sails in

* 26–35 knots or 48–64 kilometers per hour.
† 50 knots or 97 kilometers per hour.

> the harbor like a spotlight. The sea was a deep emerald green. It was fairly flat at first, but as the wind increased, the tops of the breaking waves were also a brilliant white.
>
> Just before reaching the mooring, what appeared to be a prominent tower of very dense gray mist or fog seemed to grow out of the water, rising over 100 feet* into the air it seemed, while other smaller towers came in and out of existence. Within minutes, the rain was teaming down in sheets around me and I could see nothing in the torrential downpour. The wind rose alarmingly, not in gusts, but steadily increasing in strength, accompanied by terrific lightning and thunder. Fortunately, I was able to reach my mooring whereupon I dropped sail and rode out the storm in considerable discomfort.

Bill Ericson says that engine trouble had developed aboard his boat and her sails had been slatting in a flat calm for about an hour in the open Atlantic waters off Gloucester Harbor.

> All of a sudden, I saw a great white wall of water in the air about 20 feet* tall with something in the middle that looked like a round chimney coming toward us. I tried to roll a very deep reef in the sail and had all but three feet† rolled up when it hit. The air seemed full of water. I learned later that the chimney-like thing was a waterspout which luckily passed us to port.
>
> The sea was stirred up and was a very light green color and was frothy here and there. When the wall of water hit, the boat heeled sharply and the wind was terrific. The gust must have been at least 60 knots because the sail track was ripped off nearly a third of the way up the mast and badly twisted even though there was only three feet of sail up at the time.
>
> Soon we were engulfed in an unusually heavy rain squall and the seas were quite rough. The violent wind, however, only lasted about 15 minutes, but the rain continued a while longer.

Reporting Waterspout Activity Sea lore, which seems to have an answer for almost everything that happens at sea, suggests that a cannonball shot into the center of a waterspout will cause it to dissipate. As modern technology does not seem to bear this out, mariners of today will benefit by continued research on waterspout activity. You can further this research by reporting any sightings and incidents to the National Weather Service office in your area.

---

* 30 meters.
* 6 meters.
† 1 meter.

In an article in *Weatherwise* magazine, Dr. Golden asks for help:

> Only through vigilant, organized observing programs can the true waterspout frequency and distribution be ascertained. The interested meteorologist and novice alike are therefore urged to report all funnel cloud sightings to the local Weather Service Office and, if possible, photographically document the event. *Well-documented* waterspout photography has been and will continue to be an invaluable tool in both statistical and kinematical structural studies of the phenomenon. We strongly urge that in particular all photography, both motion picture and still, be documented *at the time it is taken:* write down the date, time, location of the photographer, direction(s) photographed and brief description of the phenomenon.

Television and AM-Radio Funnel Detection Bob Synder, owner of Snyder Oceanography Services, Jupiter, Fla., offers the following methods for using television and radio to detect nearby tornado or waterspout activity.

With a television set, he says to detune the brightness on the high VHF frequency of channel 13 by turning the control so that the picture disappears and the screen is almost black. Switch the set to channel 2 and turn the volume down so the crackling noise will not startle you. Up to about 17 to 18 nautical miles (32 kilometers) away, the electrical discharge from a funnel will cause the video portion of the set to produce a steady white glow on color as well as black-and-white sets. Should you have cable TV, however, the set could react to a storm near either a microwave relay antenna or your own outside antenna. Therefore, you should detach the cable connection and use a "rabbit ears" antenna instead. It should be noted that on a set tuned for funnel detection, lightning will show up as only momentary flashes or streaks.

As mentioned in the AM and FM Radio Entertainment Broadcasts section of Chapter 5, when an AM radio is tuned to the low end of the dial, around 550 kHz, it will generate intermittent static during an electrical storm. In this same frequency range a funnel event in the area will cause continuous static.

PART II

Researching Regional Weather

Spirited yarns of wind and wave are bandied about at shoreside functions of today as frequently and intensely as they were at the whalers' gams of old. The adventurous nature that yearns for the unknown is still alive and shared by all who set out on the water, even though more do it now for recreation than for livelihood.

All over the United States, the interest in all forms of recreational boating has increased a thousandfold in recent years. Charter services are springing up almost overnight in ever increasing numbers as boaters gratify their desires for boating adventures in new areas. Those, however, who set out on unfamiliar waters without having done their weather research on the area ahead of time are wearing mental blindfolds and have set courses for disaster.

Previously, those who wished to do their weather homework for boating areas around the country have not had an easy time of it, but now Part II of this book has the situation well in hand. In it we have put together a brew of weather information for the waters of the Great Lakes and along the United States coasts of the Atlantic Ocean, the Gulf of Mexico, and the Pacific Ocean, and we have seasoned it with tidbits of local knowledge.

Before embarking on unfamiliar waters, however, it is essential to supplement the information in the following chapters with that in certain U.S. government publications. Following are the names of several that we feel all boaters should be familiar with. For ease of ordering we include pertinent addresses.

U.S. COAST PILOTS AND GREAT LAKES PILOT

NOAA publishes a series of eight books of coastal and intracoastal waters as well as the one covering the area of the Great Lakes. Each of these books gives a comprehensive presentation of weather conditions likely to be encountered in the particular section of the United States it covers as well as a dissertation on the navigational features of the area.

These books are published annually and the prices vary. They may be purchased from an authorized chart agent or ordered from NOAA. Order *U.S. Coast Pilots* and *Great Lakes Pilot* from:

Distribution Division (C 44), National Ocean Survey, Riverdale, Maryland 20840

The numbers of the *U.S. Coast Pilots* and the areas they cover are:

Atlantic Coast:

No. 1—Eastport to Cape Cod

No. 2—Cape Cod to Sandy Hook

No. 3—Sandy Hook to Cape Henry

No. 4—Cape Henry to Key West

No. 5—Gulf of Mexico, Puerto Rico, and Virgin Islands

Great Lakes:

No. 6—Great Lakes

Pacific Coast:

No. 7—California, Oregon, Washington, and Hawaii

Alaska:

No. 8—Dixon Entrance to Cape Spencer

No. 9—Cape Spencer to Beaufort Sea

Most of the weather broadcasts available for these sections of the United States are given on the *Marine Weather Services Charts* although the regional divisions are different on these charts. For further details and information for ordering *Marine Weather Services Charts* see Chapter 4, page 58.

TIDAL CURRENT TABLES

These tables are published in two volumes: Atlantic Coast of North America and Pacific Coast of North America.

Tidal currents are the horizontal movements of water into and out of harbors and other indentations along a coast. Winds can increase the predicted strength of a current when it is blowing in the same direction of the set (the direction toward which the water is flowing). When a strong wind blows against the set of a strong current, seas can increase alarmingly in height and frequency. Passage through a narrow harbor entrance is most dangerous when the wind is blowing hard directly into it and the current is flowing at maximum speed out of it.

Daily predictions of the times of maximum current and slack water as well as the drift (speed of the water flow) and the compass direction of the set are given in the *Tidal Current Tables.* Purchase from an authorized chart agent or from:

Distribution Division (C 44), National Ocean Survey, Riverdale, Maryland 20840

TIDE TABLES: HIGH AND LOW WATER PREDICTIONS

These tables are published annually in two volumes: East Coast of North and South America and West Coast of North and South America.

Two high tides and two low tides can be expected each tidal day; a tidal day averages about 24 hours and 50 minutes. Meteorologists sometimes refer to these tides as astronomical tides because they are primarily caused by the gravitational attraction exerted by the moon and the sun on the surface of the earth. A storm tide is the combination of astronomical tide and a rise in the level of the water caused by the wind from a storm pushing the water before it.

The accuracy of these tidal predictions may be affected by certain weather conditions and, according to a statement in the *Tide Tables,* "Changes in winds and barometric conditions cause variation in sea level from day to day. In general, with onshore winds or a low barometer the heights of both the high and low waters will be higher than predicted, while with offshore winds or a high barometer they will be lower."

The mean low water soundings printed on the Atlantic and Gulf of Mexico coast charts, on which the figures in these tables are based, refer to an average of the low tides for an area. The mean lower low soundings on the Pacific coast charts refer to an average of the lowest of the two daily low tides.

Tide Tables can be purchased either from your authorized chart agent or from:

Distribution Division (C 44), National Ocean Survey, Riverdale, Maryland 20840

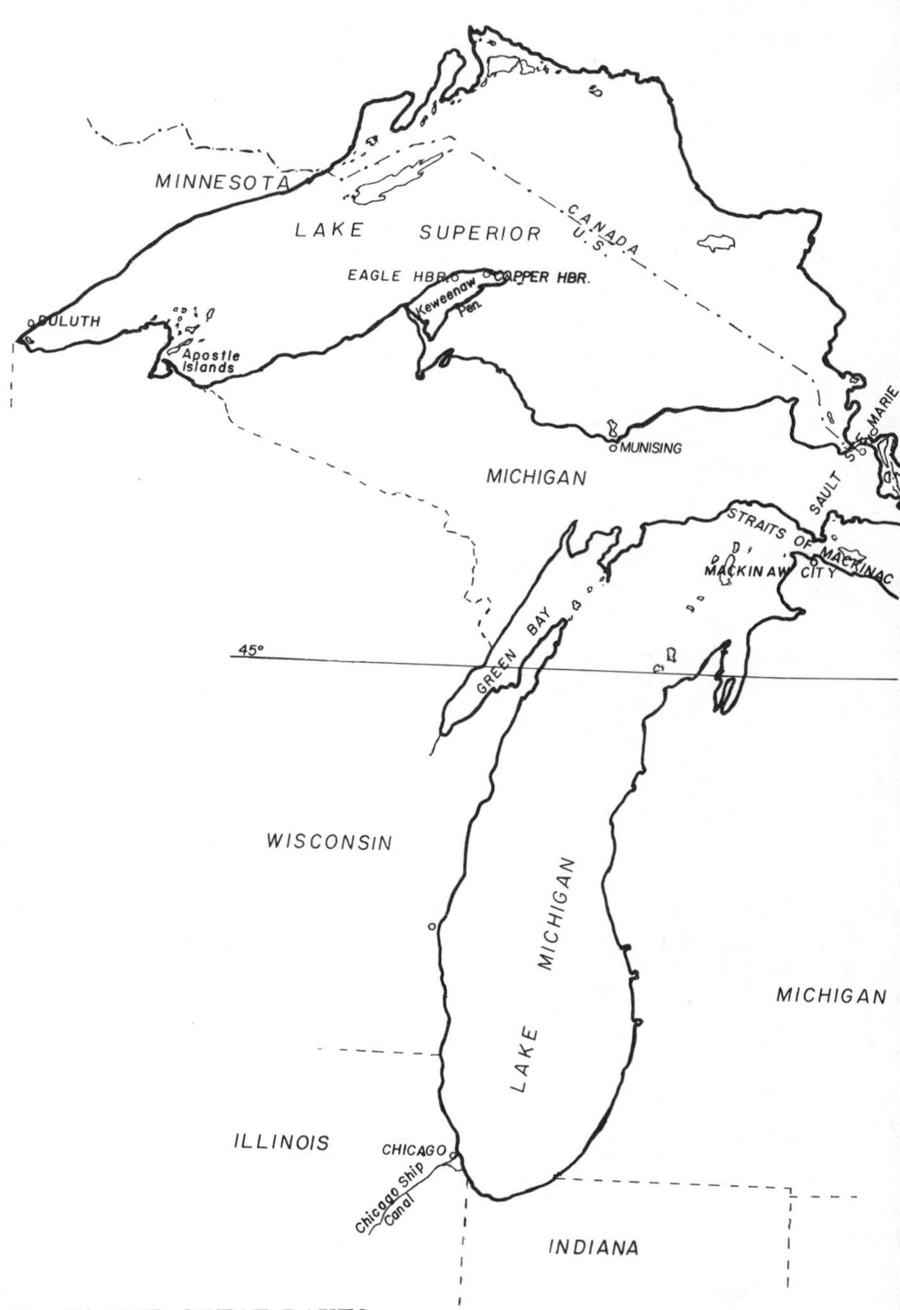

Fig. 15. THE GREAT LAKES

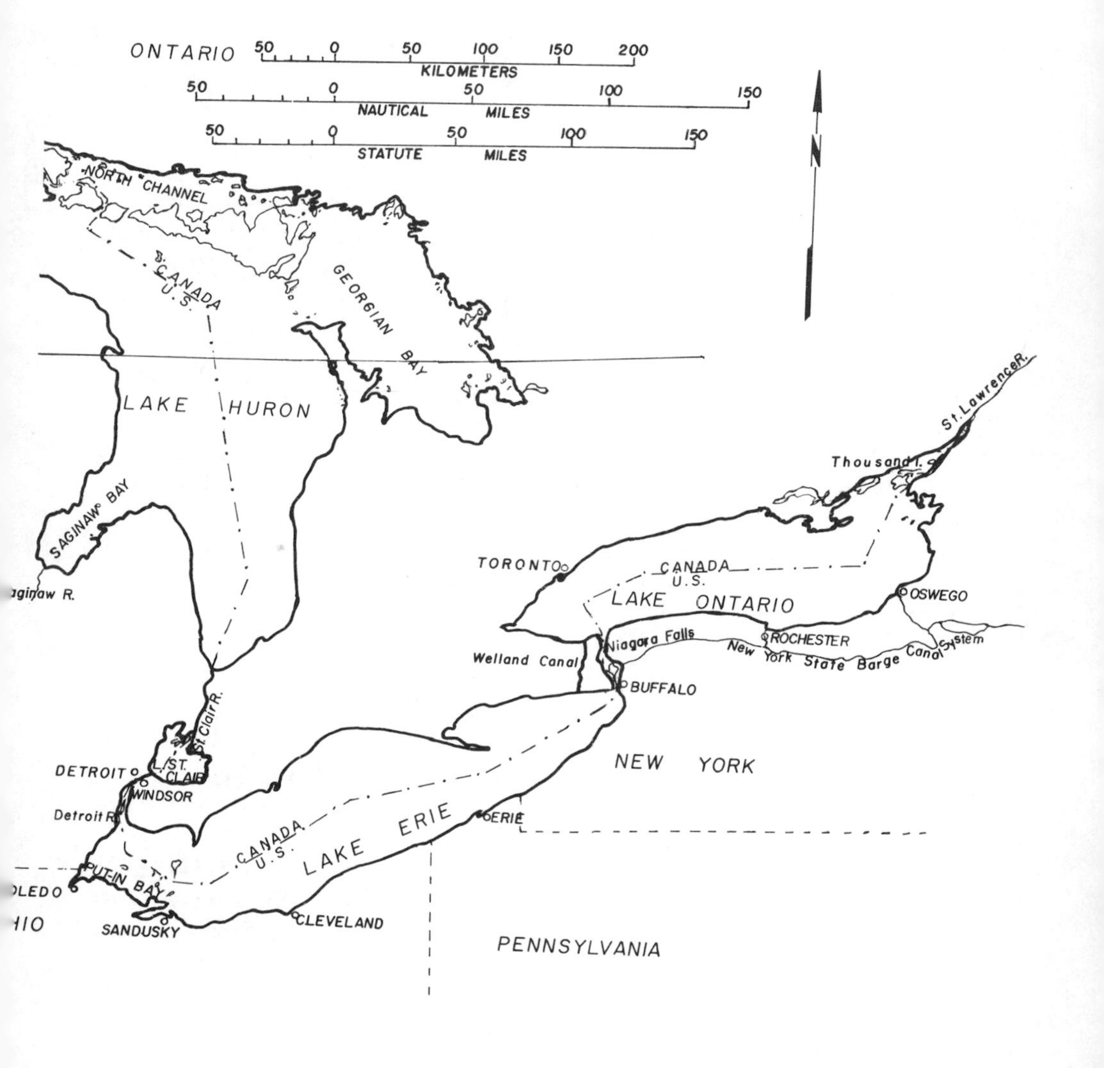
ONTARIO
50 0 50 100 150 200
KILOMETERS
50 0 50 100 150
NAUTICAL MILES
50 0 50 100 150
STATUTE MILES
N
NORTH CHANNEL
CANADA
U.S.
GEORGIAN BAY
LAKE HURON
SAGINAW BAY
Saginaw R.
St. Clair R.
L. ST. CLAIR
DETROIT
WINDSOR
Detroit R.
PUT-IN BAY
TOLEDO
OHIO
SANDUSKY
CLEVELAND
CANADA
U.S.
LAKE ERIE
ERIE
PENNSYLVANIA
NEW YORK
BUFFALO
Welland Canal
Niagara Falls
TORONTO
CANADA
U.S.
LAKE ONTARIO
ROCHESTER
OSWEGO
New York State Barge Canal System
Thousand I.
St. Lawrence R.

CHAPTER 8

Great Lakes Weather

The five Great Lakes join together to form the largest body of fresh water on earth, a vast landlocked sea covering approximately 71,400 square nautical miles (245 000 square kilometers) within the borders of the United States and Canada. The boundary between the countries in this area is a meandering 843-nautical-mile-long (1561 kilometers) line that more or less bisects Lake Superior, Lake Huron, Lake Erie, and Lake Ontario. Only Lake Michigan is wholly within the United States.

The rocks in and around the lakes are a geologic library containing remarkable fossils of plant and animal life of ages past that indicate that the formation of the lakes took thousands of years. Their present size and shape, however, are attributed to the last glacier, which bulldozed the basins as it advanced across the region and filled them as it melted during its retreat northward three to five thousand years ago.

The lakes have been used as highways for transportation and commercial ventures from the time men first settled along their shores. As technology advanced, fur-laden canoes were replaced by sailing vessels of every description. Eventually, these too gave way, as far as commercial use was concerned, to steam vessels especially constructed for lake travel.

The impetus for the construction of these steamers, or The Lakers, as they were called, was provided in the middle 1800s by the vast numbers of immigrants flocking to the area, a few to settle there, but most bound for the west. To facilitate the increased demand for travel on the lakes, connecting waterways and canals were constructed where necessary. Severe storms took their toll of the early vessels, usually during the months of October and November when there was the temptation to make one more trip before suspending operations until spring because of ice. Tales of disasters involving great loss of life and shipping make up the bulk of the lore of the lakes.

Today, a combination of comprehensive, up-to-the-minute weather reports, predetermined shipping lanes, and technological advances in ship construction make the transportation of millions of tons of cargo in heavily laden barges and deep-draft freighters possible for about eight months of the year with relatively few losses as compared to the misfortunes of the past. The use of icebreakers helps to extend the number of months in which some ships can operate.

U.S. AND CANADIAN WEATHER WARNINGS

The consensus among many widely traveled commercial and recreational boatmen is that the cooperative U.S. and Canadian weather warning system for the Great Lakes is outstanding.

Weather reports from ships often alert the National Weather Service to unusual weather conditions. Every effort is made to broadcast information from these sources as quickly as possible. According to the National Weather Service forecast office in Cleveland, weather reports from each ship are now received simultaneously at each marine radio station, at Coast Guard headquarters, and at all the Weather Service forecast offices in the region—Milwaukee, Chicago, Detroit, Cleveland, and Buffalo in the United States, and Toronto in Canada. As a result, a warning may be issued and broadcast widely almost immediately after receiving any special report from a ship that has encountered gale or storm-force winds or other hazardous conditions.

MAFOR

MAFOR is the name used for the combination number-coded/plain-language weather forecasts broadcast over the radiotelephone. The key used for decoding MAFOR transmissions appears on the two Great Lakes *Marine Weather Services Charts* and in the *Worldwide Marine Weather Broadcasts.*

MAFOR is intended primarily as a service for the large commercial ships plying the open waters of the lakes and through the St. Lawrence Seaway, but pleasure boatmen have also found these coded weather forecasts extremely helpful. A friend with more than 35 years of experience in commercial shipping on the lakes says, "You can get a lot of information from numbers that might be missed in word transmission, but you must mark down the message clearly, always writing down each figure."

When the spoken word "MAFOR" is used on a radiotelephone frequency, it is an alert for a marine weather forecast. Then a number-coded part of the forecast follows in this order:

1. Four numbers to indicate the date and time of the beginning of the forecast period
2. The name of the body of water (this is given in plain language)
3. One or more sets of five numbers to represent the expected wind direction, wind speed, and weather conditions for different forecast periods for the particular body of water named (Each set of numbers is preceded by the number 1, which is the identifying number for the Great Lakes and the St. Lawrence River area and is included in the transmissions by international custom)

In the remainder of the MAFOR broadcast, sea conditions are always given in plain language as is a brief synopsis of weather conditions over Canada and the United States that might affect the area. The synopsis is based on the most recent observations by the National Weather Service in the United States and the Aids to Navigation Division of the Department of Transport in Canada.

LAWEB

LAWEB is another weather-data broadcasting system used on the Great Lakes. The Service is best described in the following excerpt from a 1977 *Marine Weather Services Chart:*

> U.S. and Canadian ship and coastal observations are broadcast by radiotelephone every 6 hours during the navigational season . . . . The LAWEB contains plain language reports of wind direction and speed (in knots) and wave heights from shore stations in the Lake Region and ships underway on the lakes. Ship's positions are given in distance in miles and direction from well-known landmarks; visibility and weather are included in all reports when visibility is less than ⅝ of a mile. Observations are taken 1½ hours prior to the times of broadcast.

The weather warnings and forecasts become more meaningful once a boatman is aware of the types of weather conditions that are likely to occur on the Great Lakes.

STORMS

Perhaps more to keep this summertime boating paradise to themselves than for any other reason, recreational lake boatmen are likely to dwell on "storms I have survived" stories when recounting their lake boating experiences, so that to someone from another part of the world, bad weather seems to be the rule. Although the Great Lakes have a justly deserved reputation for exceptionally severe and long-term storm activity, statistically a general pattern of good weather appears to prevail during June, July, and August on Lake Erie, Lake Ontario, and on the southern halves of Lake Michigan and Lake Huron. This is also true during July and August on Lake Superior and the northern areas of Lake Michigan and Lake Huron. Severe storms do occur from time to time on the lakes during the summer months, but generally they are of such short duration that the danger from wave activity is considerably less than at other times of the year.

Summer adventures or misadventures with weather are likely to be caused by carelessness. The old-timers speculate that this usual summer pattern of good weather lulls the inexperienced into a false sense of security. They warn of the dangers of either disregarding visual warning signs of worsening weather or ignoring radio broadcasts of weather bulletins. The unwary boater is fair game for sudden summer thundersqualls.

Local thunderstorms, often quite intense, form over land near the coasts in the afternoon and evening when there is a high degree of moisture in the air and the surface of the land has become heated by the summer sun.

The most violent storms are usually associated with thunderstorm activity that is part of a strong frontal system. These storms can be particularly rigorous in the spring or early summer when a full-fledged cold front, still packing the wallop of the winter weather pattern, blasts down from Canada and collides with a moist warm air mass pushing up from the south. Therefore, as tornadoes are frequently associated with these storms, this is the season when the Great Lakes have the greatest number of tornadic waterspouts. Many waterspout sightings have been reported over the years and lake historians speculate that the mysterious disappearance of some boats may have been caused by this weather phenomenon.

WAVES

Mariners on these waters have great respect for the waves that can build up, particularly on the long, open stretches of the lakes.

There are even reports of waves during the stormy fall and winter months with heights rivaling storm waves on the oceans.

During the boating season, the recreational boatman can anticipate the likelihood of a steep, bone-jarring chop occurring on any of the lakes when a fairly strong, steady wind is combined with a long fetch. The shallower the body of water, the more pronounced the chop will be.

STORM SURGES

Storm surges, sometimes referred to locally as wind-driven tides, cause changes in the water levels of the lakes. (Astronomical tides, as experienced on ocean waters, are nearly zero on the lakes.) Moderate to strong winds, blowing for a sustained period of time from one direction, or a squall line, with its accompanying changes of barometric pressure, can result in a pileup of wind-driven water along any shore exposed to a long fetch. Shoaling occurs along the opposite shore. Storm surge can continue for several hours or only for a few minutes, depending on the severity or duration of a storm.

Wind-driven water buildup with its counteraction across a lake is greater where there is a long fetch. It is further aggravated when the normal subsurface return flow of water is hindered by shallow depths. Lake Erie, which is the shallowest of the Great Lakes, has a long west-to-east fetch between Toledo, Ohio, and Buffalo, N.Y.; a maximum difference of 4 meters (14 feet) in water level has been recorded between these two cities as the result of a storm surge.

SEICHE

As long as a strong wind continues to push water onto the shore of a lake or bay, it will retard the ability of the water to return to its normal state. When this restraint is removed suddenly, such as when the storm abates and the velocity of the wind drops, a new phenomenon called seiche (sayshh) enters the picture. The action of a seiche in a lake is similar to what happens to the water in a dishpan when the pan is jostled. The liberated water tends to rock from one side to the other until equilibrium is restored.

Seiche is defined by the National Weather Service as "a continuation of a water level disturbance after the external forces causing the disturbance have ceased to act." As seiche waves can take a considerable amount of time to cross a large expanse of water, they are likely to arrive on a shore long after the storm is over, seemingly from out of nowhere. Boatmen, therefore, should be alert for seiche waves after any severe or prolonged blow.

Seiche waves, as with all waves, are highest in shallow bodies of water where free movement is restricted by friction with the bottom. On Lake Erie, for example, once the initial buildup of water at Buffalo is released, a seiche wave will take about 14 hours to make a round trip from Buffalo to Toledo and back again. This round trip may recur several times. After a severe storm, when the water has built up 2.4 meters (eight feet) or more at Buffalo, as much as three days may be required for the lake to return to its normal level.

National Weather Service warnings of storm surge conditions and possible seiches are issued in marine warning bulletins on radio and television. Local police officials in areas likely to be affected are contacted directly.

WATER LEVEL FLUCTUATIONS

The water levels of the Great Lakes tend to fluctuate seasonally with higher water in the summer and lower water in the winter. The water levels can also vary from year to year so that some years the lakes seem very full and other years residents find their piers too high off the water for boating convenience.

Lake water levels change according to the amount of rain and snow falling directly on the surfaces, runoff from countless small rivers, streams, and brooks draining land areas in Canada and the United States, seepage from underground water sources, flow of excess water from a higher lake to one below, and surface evaporation.

Information concerning the existing and expected levels of each of the Great Lakes and Lake St. Clair (between Lake Huron and Lake Erie) is provided from May through October by monthly supplements to the *Great Lakes Pilot*.* This information is also available in the monthly *Bulletin of Lake Levels** and is often published in newspapers and reported over radio and television stations. Measurements are given for rise and fall above or below the low-water datum depth figures written on navigational charts.

FOG

Fog on the Great Lakes is more prevalent in the northern regions where the water stays colder than in the southern sections of the lakes.

---

* Distribution Division (C 44), National Ocean Survey, Riverdale, Maryland 20840.

It occurs most frequently in the spring due to the tendency of the water temperature to lag behind the temperature of the surrounding land during a warming trend. By summer, however, fog is not a serious problem for boating. It occurs primarily along the shoreline and usually burns off by mid-morning.

We have been told by an experienced lake boatman that it is almost impossible to predict the formation of advection fog, the type that occurs when hot and humid air from the land is blown over the cold water and is chilled immediately to the dew point. However, a Great Lakes National Weather Service meteorologist disputes this. He says. "It's easy if you know how. A simple psychrometer will give you the dew point. Anytime this is higher than the water temperature, fog will form offshore."

Advection fog sometimes produces a long dense ribbon of fog one-half to one nautical mile from shore, and other times dense fog develops along the shoreline. The offshore fog bank occurs more frequently, and, when it is present, the visibility near shore and inland is likely to be clear.

REFRACTION

Veteran seamen on the Great Lakes often comment on the beautiful and mystifying mirages that so frequently occur there, especially on the cold waters of Lake Superior. One of the best descriptions of the phenomenon is the following account from the *Mariners Weather Log*, NOAA's Environmental Data Service magazine, written by Richard W. Schwerdt. Mr. Schwerdt describes his impressions of these optical illustions as viewed from the deck of the ore ship *Weir* en route to the Keneewaw Peninsula on Lake Superior.

> As we rounded the peninsula, just a few miles off the heavily forested coast, we could pick out the seaside towns of Copper Harbor and Eagle Harbor from the *Weir*. The clean air and the sunlit waters played tricks on your eyes at that time of day. A number of eastbound boats (located between the sun and us) were distorted. Sometimes they appeared as a solid block (stooping) rather than higher at both ends as ore boats should look and at other times the stern appeared to be three or four times higher than the pilot-house (towering). Stooping and towering are distortions caused by the atmospheric refraction of light and are only noted when there is a strong temperature inversion—in this case, much warmer air over cooler air near the surface of the cold lake.

CHARACTERISTICS OF INDIVIDUAL GREAT LAKES

Each lake has special qualities that are not necessarily shared in common by all. These result from the particular configuration of a particular lake, its depth, or its geographical location.

A ship leaving Lake Superior at the northwestern end of the chain of the lakes will journey, via the deep-water channels of the St. Lawrence Seaway and the majestic St. Lawrence River, approximately 1,500 nautical miles to the sea, traversing Lake Huron, Lake Erie, and Lake Ontario along the way. The vessel begins its passage at a height of 180 meters (600 feet) above sea level and by the time its bow slices into the steely Atlantic Ocean swells, it will have stairstepped down all but 12 meters (38 feet) by means of ingeniously engineered locks. Most of the natural 12-meter descent occurs gradually in the St. Lawrence River.

Lake Huron is reached from Lake Superior via the Corps of Engineers locks between Sault Ste. Marie, Mich., and Sault Ste. Marie, Canada. Passage from Lake Huron into Lake Michigan is made through the free-flowing Straits of Mackinac situated to the north of Mackinaw City, Mich. At the southern toe of Lake Michigan, the connecting Chicago Sanitary and Ship Canal, the Illinois Waterway, and the mighty Mississippi River provide a commercially profitable towboat and barge route to the Gulf of Mexico.

Lake Erie is reached from Lake Huron through the St. Clair River, Lake St. Clair, and the Detroit River, Detroit, Mich., and Windsor, Ontario, face each other across these greatly traveled waters.

Two outlets, the Niagara River, with its world famous falls, and the Welland Ship Canal connect Lake Erie to the easternmost of the sister lakes, Lake Ontario. The great difference in water level between Lake Erie and Lake Ontario, the latter being 100 meters (327 feet) lower, accounts for the spectacular Niagara Falls.

The deep-water channels with controlled depths of at least 8 meters (27 feet) and the enormous locks of the joint U.S. and Canadian project, the St. Lawrence Seaway, opened the heartland of the two countries to international shipping in 1959. For the first time, modern vessels that traversed the oceans of the world could enter this vast inland sea as well. Previously, the size of ships entering the Great Lakes from the Atlantic Ocean was limited by the controlling 4 meter (14 foot) depth of the St. Lawrence River or the 2.7 meter (9 foot) depth of the New York State Barge Canal, which links the lakes to this ocean via the Hudson River.

The following dimensions in nautical miles have been adapted from information that appeared in a government publication entitled *Data on the Great Lakes System:*

	Superior	*Michigan*	*Huron*	*Erie*	*Ontario*
Length in nautical miles	304	267	179	209	168
Breadth in nautical miles	139	103[a]	159[b]	50	46
Length of coastline including islands in nautical miles	2,590	1,443	2,763[c]	744	631
Areas in square nautical miles:					
Water Surface,					
United States	15,600[d]	16,800[e]	6,870[f]	3,760	2,610[g]
Canada	8,380[d]	——	10,500[f]	3,720	3,010[g]
Average Depth:					
meters	149	85	59	19	86
feet	489	279	195	62	283

a. Measured at wide point through Green Bay
b. Measured at wide point through Georgian Bay
c. Includes Georgian Bay and North Channel
d. Including St. Marys River above Brush Point
e. Lake Michigan including Green Bay
f. Including St. Marys River below Brush Point, North Channel, and Georgian Bay
g. Lake Ontario including Niagara River and St. Lawrence River above Iroquois Dam

Lake Superior Lake Superior, situated at the northwestern end of the chain, is best described with superlatives, as it is the largest, longest, deepest, most northerly, coldest, clearest, and cleanest of the Great Lakes. It is second only to Lake Huron in breadth. It is bordered by Ontario, Canada, to the north and in the United States by Michigan and Wisconsin to the south and Minnesota to the west.

Those who cruise there tend to become almost lyrical when describing the magnificent north shore with its rugged, tree-studded cliffs and deep bays. Waves, wind, and rain have sculptured the red sandstone cliffs near Munising, Mich., on the south shore into natural works of art called the Pictured Rocks, to mention only one of nature's masterpieces to be found on the lake.

The water temperature on this northernmost lake rarely ever reaches 13°C (55°F). The harbors can be icebound until the end of May, and beds of snow sometimes remain on the banks through the month of June.

During the summer, fog is often prevalent in the northwesterly portion where the waters stay the coldest, while it is less likely to occur along the southern shore during this period. Since such a long fetch is possible from all directions, substantial seas can build up whenever there is a strong wind. The great depth of the lake, however, lessens the danger of seiche wave action once a storm has abated, although the danger is still present in shallow bays such as in Keweenaw Bay inside Keweenaw Peninsula where a seiche-produced water rise of 1.8 meters (6 feet) has been reported.

Recreational boaters who contemplate cruising the lake at any time except during July and August should consider the following terse warning for Lake Superior that appears in the *Great Lakes Pilot:* "In the spring and fall the lake is foggy, stormy, and dangerous. Late in the navigation season, ice accumulation on ships is sometimes a problem."

Lake Michigan Lake Michigan is the only one of the lakes that is totally contained within the borders of the United States. Four states share its shores, with Michigan to the east and north, Indiana and Illinois to the south, and Wisconsin to the west.

In the summer, Lake Michigan is a veritable mecca for boating enthusiasts. There is a large selection of excellent marinas and either man-made or natural anchorages along both shores of this lake. The Army Corps of Engineers in cooperation with the Michigan State Government is currently engaged in a project that will enhance recreational boating by providing harbors about 12 nautical miles apart along its shores. (Similar programs are underway in Lake Huron and Lake Superior.) In addition, there are many small lakes and inlets along the shore that offer protection to the outboarder and day sailor.

A retired Coast Guard commander who taught weather courses to many groups in Michigan, including the Coast Guard Auxiliary and Mariner Scout troops, says that the entire eastern shore of the lake is one long and dangerous lee shore in the nor'westerlies that can be expected from time to time. Lower Lake Michigan is noted for summer thundersquall lines that move from west to east across the lake creating a water wave along their leading edge. The water disturbance is seiche-producing. When this phenomenon occurs, it is particularly noticeable because it affects Chicago's highly popular lake shore.

According to a National Weather Service newsletter:

> All seiche-producing squall lines move from the west across (lower) Lake Michigan, so the windy, rainy weather moves through Chicago first, without significant effect on lake levels. The line moves across the lake, creates the wave, and causes the water level to rise first on the eastern shore with a surge at about the same time as the thunderstorms. The coincidence of the two events gives some protection to the eastern shore, as the approaching thunderstorm is likely to cause people to seek shelter before the surge hits the beach. As the thunderstorms move into Michigan, the surge then reflects from the eastern shore back to Chicago as a true seiche.

The National Weather Service in Chicago has issued seiche forecasts for lower Lake Michigan since 1954. It is very difficult to predict the height of a seiche wave, but they have had remarkable success predicting the time of the arrival.

The commander cites possible weather conditions for two areas on the western shore in particular, Milwaukee Harbor and Green Bay, Wis. He says that in the Milwaukee Harbor area, the backwash from wind-driven tides during a nor'easter or easterly storm creates steep waves as far out as two nautical miles from the shore. The sea becomes so confused that it can be a wild ride into the harbor. He remembers many accidents that occurred in these conditions, including some sailboats that were dismasted. On one occasion while aboard a large tug in the harbor, he recalls that it was only blowing about 25 knots, but the seas had built up so high that one minute he could see downtown Milwaukee and the next he was in a water-surrounded hole. When it is hot ashore in the Green Bay area, he continues, sudden severe thunderstorm activity and fog are likely to occur.

The *Great Lakes Pilot* describes weather conditions on Lake Michigan in the following way:

> The long axis of the lake, north and south, aggravates the effects of northerly and southerly strong winds and gales, affording an extended fetch for the propagation of dangerous seas, and creating strong currents and hazardous conditions at the harbor entrances. It is of importance to note that the stronger winds of the fall, winter, and early spring are usually from the westerly quadrants, making entrance to the restricted harbor channels of the east shore especially difficult, and sometimes increasing the dangers by the formation of storm bars. In the spring, navigation in the northerly part of the lake is rendered perilous by fog and by shifting ice fields, particularly in the region of the Straits of Mackinac.

The Ericsons' friends Jim and Dee Dilworth, who cruise and race their small auxiliary in the summertime on the northern part of Lake Michigan, say that the water temperature is usually chilly for swimming until August. "The lake," they add, "can be very windy and choppy, but it also has been so calm that we have had to use the outboard. It can be like glass and many times it is foggy. The lake is changeable like the ocean and big seas can rise from strong west winds. During the summer, the prevailing winds are from the southwest to the west. Most nights find the lake calm as the wind comes up about 8 or 9 A.M. and goes down about 5 or 6 P.M."

Lake Huron Many cruising enthusiasts agree that some of the most magnificent scenery and solitary, peaceful anchorages on all of the Great Lakes can be found on island-studded Lake Huron, particularly in the Canadian North Channel and Georgian Bay areas.

Sallie and Sam Townsend listened to many tales of cruising on Lake Huron from their friends Louise and Fritz Garber when the couples were cruising in the Bahamas several years ago in their own boats. At first, the Garbers found it difficult to accept the crowded marina life along the eastern coast of the United States and even the increasingly crowded harbors of some of the Bahama Islands. They yearned, instead, for their home cruising waters of Lake Huron and especially for Georgian Bay where night after night they could find their own private anchorage in a beautiful secluded spot with no other boats within sight or sound. The Garbers have since checked in with the news that those halcyon days on the lake are fast becoming a memory; too many others have joined their sport!

Although Lake Huron is the second-largest of the Great Lakes, it is similar to Lake Michigan in size and water depth; but the population settlement along the shores is sparse in comparison. The lake is important as a commercial artery because it is a link between the upper and lower lakes. Its waters lap the shores of Michigan in the United States to the west and the shores of Ontario, Canada, to the north and east.

The winds and waves that can occur on Lake Huron are described in the *Great Lakes Pilot* as follows:

> The configuration of Lake Huron is such that winds from any quarter may stir up considerable wave action. The long fetch of strong northeast and east winds over the waters from Georgian Bay to the Michigan shore develop high seas which run athwart north and south courses of traffic through the lake. Winds from the south may be dangerous, especially in the northern converging shores of

> this lake and north winds are dangerous in the pointed southern configuration of the lake, especially near the outlet of it . . . . The highest velocity as observed on anemometer-equipped vessels while more than 6 miles offshore was 95 knots (109 statute miles per hour) on 6 August 1965 . . . . [This wind speed, as far as we can ascertain, is the highest recorded on any of the Great Lakes.]
>
> Wind driven tide is materially augmented in bays . . . where the impelled water is concentrated in a restricted space by the converging shores, especially if coupled with a gradual sloping inshore bottom which reduces the depth and checks the reverse flow via lower currents. The latter condition is very pronounced at the mouth of the Saginaw River.

Lake Erie Although Lake Erie is next to the smallest of the lakes, three of the Great Lakes' major ports are situated along its shores: Cleveland, Toledo, and Buffalo. The lake is bordered to the north by Ontario, Canada, and to the south and east by Ohio, Pennsylvania, and New York in the United States. The state of Michigan is at its western end.

The warm waters of this southernmost lake encourage vacationers and residents alike to engage in all types of water sports throughout the summer months. The serpent in this Garden of Eden, however, is that this lake has the justly deserved reputation of being the stormiest and most dangerous of all the Great Lakes. Because of its shallow depth (average depth 19 meters or 62 feet) and 209-nautical-mile-long southwest-to-northeast fetch, its waters are often choppy even when the winds are fairly light. During a storm, steep waves build up quickly and storm surges and seiches occur. The water level can rise appreciably at the northeastern tip of the lake when a storm is from the southwest. A nor'easter can cause water buildup from Sandusky to the Detroit River entrance on the western end of the lake, and particularly at Toledo.

The area is noted for its sudden late-afternoon thunderstorms in the summer. A few years ago on a sunny, hot July afternoon, Virginia observed dark, ominous storm clouds building up in the northwest while she was swimming off the beautiful sandy beaches on Presque Isle Peninsula in Erie, Pa. The lifeguards quickly ordered everyone out of the water and cleared the beach. Within minutes, it seemed, the wind was blowing strongly, driving the rain before it with gusts of 30 knots or over. The waters of the lake were churned into a froth of high, confused waves.

Friends who cruise on this lake recommend that the many islands of the Put-In-Bay area in the southwestern part provide protection in bad weather with good holding-ground anchorages off safe shores. They also suggest that those with small boats remain in protected waters such as the Detroit River or the Niagara River whenever the weather looks chancy. Persons who venture from the shore at all are advised to keep alert for visual weather signs and to listen to the National Weather Service broadcasts frequently.

Lake Ontario Lake Ontario is the smallest of the Great Lakes and is situated between Ontario, Canada, to the west and north and New York State to the east and south. The important Canadian city of Toronto is situated on the northern shore at the western end of the lake.

The most picturesque area for cruising is near the eastern end of the lake where its 26-nautical-mile width is crossed by a chain of five islands. A friend of the Townsends who lives in Toronto and cruises extensively aboard his power boat claims that the islands in this part of the lake offer some of the best cruising grounds he has ever seen. He also adds that it is only a short jaunt to the eastern tip and down the St. Lawrence River to reach the fabulous and storied Thousand Islands.

Lake Ontario is subject to frequent summertime thundersquall activity. Everyone should be alert for visual weather signs and should listen frequently to National Weather Service broadcasts. Although there is a long west-to-east fetch, the danger from severe buildup of storm waves is considerably less on this lake than on some of the others because of its great depth (average depth 86 meters or 283 feet) in comparison to its size. Of all the lakes, little Lake Ontario is second only in depth to huge Lake Superior. Therefore, the effects of wind-driven tides are usually only pronounced in the bays or at one end or the other of the lake, depending on the direction of a storm.

Waterway acquaintances of the Townsends, who were on their way south from the Great Lakes aboard their bright red auxiliary, prefaced their remarks about Lake Ontario with the terse statement, "In the spring, stay off it!" They explained that although the lake does not freeze because it is so deep, the water is so cold after the winter that a lot of fog forms when warm air starts to flow over it in the spring. Once the water warms up to about 16°C (60°F) or so, the air becomes crystal clear and boaters find themselves in a cruising paradise. According to these informants, Lake Ontario offers some of the most relaxed and protected cruising you can find anywhere.

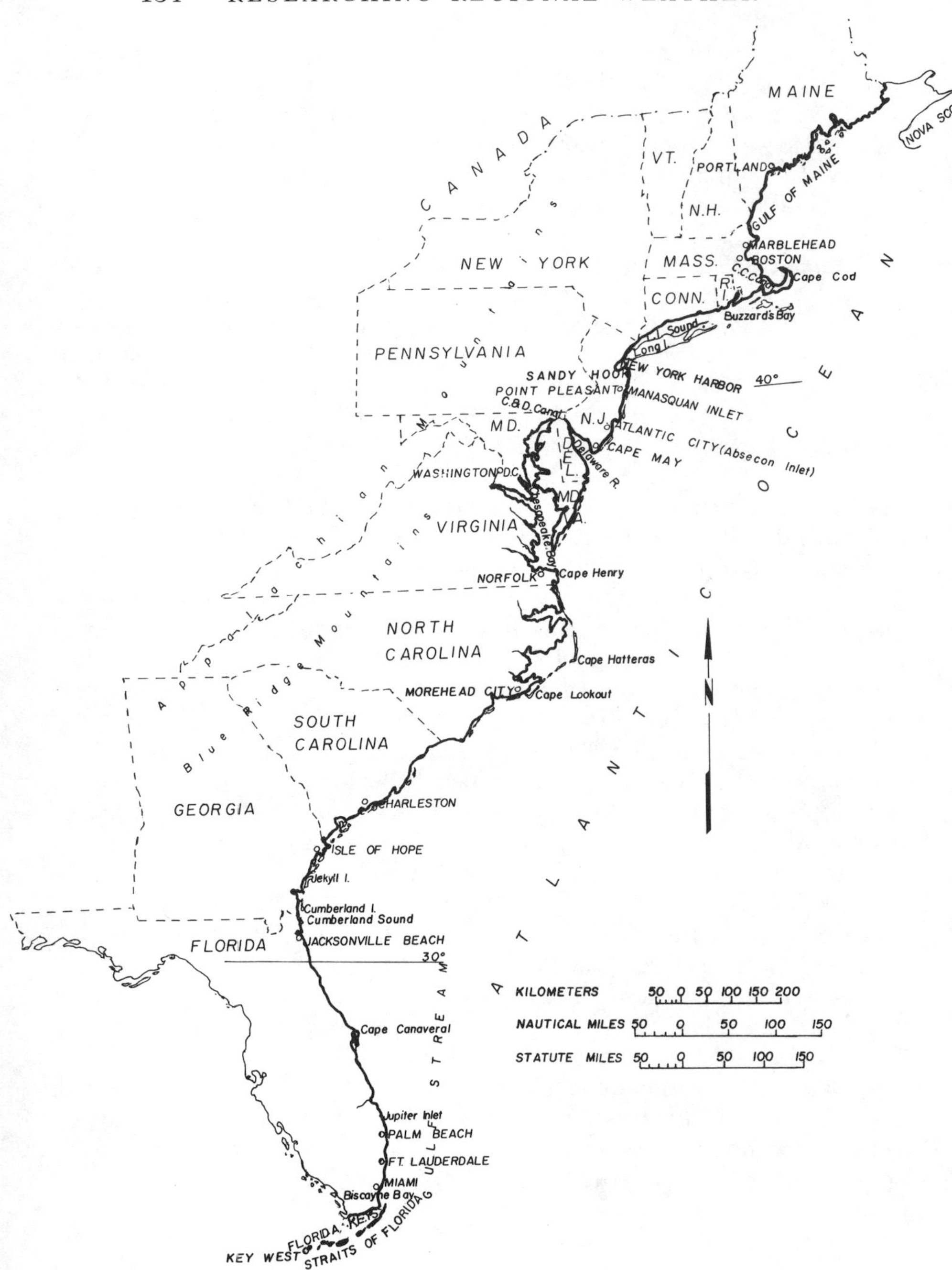

Fig. 16. THE ATLANTIC COAST

CHAPTER 9

Atlantic Coast Weather

The early history of the eastern coast of the United States is rich with tales of seafaring men who guided their vessels in search of freedom or profit through the scores of sounds, inlets, and bays along this Atlantic Ocean shore. From the days when the region was best known as the New World, caravels and galleons, frigates and clippers, whalers and coasters have swung at anchor within the many protected harbors, behind the countless wooded islands, and in natural waterways inside the miles of wind-blown, wave-lashed barrier beaches.

Harbor towns initially grew up along this coast to provide services for seafarers. Some of these towns, such as the world-famous ports of Boston, New York, and Baltimore, have continued to thrive since commercial ventures first found a foothold here. As the seafaring picture changed over the years to larger and deeper-draft ships, other coastal towns less fortunately situated were threatened with extinction as busy ports. The tremendous and comparatively recent growth of recreational boating, however, has infused them with new life. Boatyards in small harbor towns from the northern reaches of Maine to the tip of the Florida Keys once more hum with activity and local businesses flourish providing services for this new type of seafarer.

This varied boating paradise requires that all who wish to enjoy its pleasures safely must be weather-wise, both visually and intellectually. It sometimes seems that Mother Nature works overtime on this Atlantic shore to keep weather forecasters dangling from the end of their predictions, although here of all places in the country the weather mystery ought to be the least elusive. After all, one would think meteorologists here should find extended forecasting a comparatively easy task, forearmed as they are with a vast amount of information concerning weather system movements across the entire width of the North American continent. Contrary to expectation, however, there are so many influences on this coast's weather that it is hard to decide which one, if any, will insist on a majority vote in the weather forum and aggravate, weaken, or stall a usually predictable weather system.

CHANGES IN AIR MASSES

As the Atlantic coast of the United States as far south as northern Florida falls within the terrestrial wind belt of the Prevailing Westerlies, most air masses that affect this area are carried by the terrestrial winds across the continent in the usual west-to-east direction. However, they may change significantly as they are carried over different geological formations, such as large bodies of water or steep land elevations.

For example, if cold fronts from the north cross the Great Lakes or warm fronts from the south move over the Gulf of Mexico, they are likely to draw up moisture and thereby produce an increase in the precipitation associated with them by the time they reach the shore. On the other hand, when the lower edges of other fronts are impeded by the peaks of the over 2400-kilometer-long (1,500 statute mile) Appalachian Mountain chain, especially cold fronts destined for portions of the coast from Chesapeake Bay southward, their strength is usually mitigated. The Appalachian Mountain system stretches in a broad band, from scattered hills on the ocean's edge in Newfoundland and northern New England, inland in a southwestward direction, to the center of the state of Alabama.

THE BERMUDA-AZORES HIGH

In the spring, summer, and the first month or so of fall, the advancement of weather systems over different sections of the coast is sometimes blocked by the Bermuda-Azores High, a permanent high pressure area that migrates seasonally over a large portion of the North Atlantic Ocean.

Much of the high's influence on coastal weather is dependent upon its size and location. In the winter, the high has little or no effect on the weather in this region because it is comparatively small in size and strength as well as being located far out to sea southeast of the United States. Thus the forceful frontal systems from the west can drive unimpeded out over the ocean allowing their accompanying storms free rein to lash the coastal waters with violent winter fury.

The movement of the weather systems moving in from the west gradually becomes more and more restricted in the spring when this great circulatory body of air expands and moves north and westward toward the coast. The Bermuda-Azores High's influence on air mass progression is summarized from the Atlantic Coast volumes of the *U.S. Coast Pilot* as follows:

> During the spring, the Azores High affects the southern Atlantic coast when frequency of storms decreases and the weather becomes uniformly warm and humid. In the spring the middle Atlantic area usually is located outside the high-pressure circulation, however, and is still subject to the passage of frontal activity and changing air masses. Warm spells, sometimes with abundant rain, alternate with cool, dry weather.
>
> In the summer, the Bermuda-Azores High reaches its most northerly and westerly position, embracing the entire eastern seacoast within its circulation. The strength of this circulation is moderate but persistent, sufficiently so to hold back the east-movement of the continental low-pressure system. As a consequence, the daily weather along the coast may not change for several weeks at a time; it is controlled by the southerly and southwesterly winds bringing moist warm air from the Gulf. This weather is characterized by frequent instability, showers and thunderstorms, uniformly warm temperatures and high humidity, and relatively low wind speeds.
>
> In the autumn, the Bermuda-Azores High again shifts southward and eastward. This gradually gives way to the winter weather pattern, bringing increased frontal activity.

Sometimes, this high is also responsible for deflecting the projected northeastward course of a hurricane from heading out to sea after its recurvature so the destructive fury of a whirling monster storm is unleashed instead on the eastern seaboard of the United States.

OCEAN CURRENTS

Two massive ocean currents are responsible, in part, for the climate of different regions of the east coast. They are the cold Labrador Current and the warm Gulf Stream, which flow toward each other near the Atlantic shores of the United States. Where there is an infusion of cold current into the waters near the coast, lower air temperature and more frequent fog will prevail.

These two North Atlantic currents are part of a great system of currents that course through the oceans of the world like mighty rivers. Their paths, as in the case of terrestrial winds, are primarily determined by the force of the rotation of the earth, yet their flow is often diverted by land masses. Ocean currents do not reverse their set (direction toward which the water is flowing) each day as is the case of the familiar tidal currents, which are affected by the daily rise and fall of the tides.

The Labrador Current flows southward from Davis Strait between northern Canada and Greenland. The Gulf Stream sets to the northward from the Straits of Florida, which stretches for 85 nautical miles between Cuba and southern Florida.

Mariners have used these ocean rivers for centuries to speed their vessels along, often going far out of their way to ride a favorable current and thus trading the longer distance traveled for a quicker passage. This practice is especially prevalent in the Gulf Stream, for it is by far the stronger of the two currents.

For generations, seamen have marveled at the phenomenal swiftness of the Gulf Stream. Scientists generally attribute this as well as the tepidity of its temperature to the great flow of the two equatorial ocean currents that feed into it after being diverted by land from their normal westward path, which encircles each side of the equator. South of the equator, the South Equatorial Current divides at the bulge of the South American continent and much of it turns toward the northwest curbed by this coast. It is joined by its northern counterpart, the North Equatorial Current, and together these waters are forced to head north to northeast by the concave curve of Central America. A portion splits off to loop around the Gulf of Mexico. All these currents converge again in the Straits of Florida where they are joined by lesser currents as they start the race up the Atlantic coast and are known as the Gulf Stream. This newly formed great ocean river can flow with a speed upward of five knots.

In 1855, Mathew Fontaine Maury, one of the first men in the United States to study and record ocean meteorology, wrote *The Physical Geography of the Sea,* which includes the following passage:

> There is a river in the ocean: in the severest droughts it never fails, and in the mightiest floods it never overflows; its banks and its bottom are of cold water, while its current is of warm; the Gulf of Mexico is its fountain, and its mouth is in the Arctic Seas. It is the Gulf Stream.

The Gulf Stream is a visible and well-defined deep blue swath cutting through the normally brown-tinged Atlantic Ocean. The abrupt temperature change that occurs on the United States side of the stream is called the "cold wall."

The ocean currents affecting the Atlantic coast of the United States are under constant surveillance by NOAA, and their records indicate that the Gulf Stream can meet the Labrador Current anywhere between Nova Scotia and New York. This meeting place is subject to change largely because surface winds are likely to exert a great deal of influence on the speed of their passages. Wherever the main streams of these currents encounter each other, most of the cold Labrador Current sinks below the warmer Gulf Stream to continue its southward journey below it. However, some of the Labrador Current's south-flowing water surfaces close to the coast to run counter to the northward-flowing Gulf Stream.

FOG

The temperature of the air reflects the temperature of the water it travels over for any appreciable distance. Air that is blown over the waters of the Gulf Stream is warmed and its moisture content increases. When this moist warm air is then blown over the cold water of the Labrador Current, it is cooled suddenly and dense advection fog forms. A light to moderate breeze can roll this sea fog, as it is often called, in from the ocean, where it can shroud a coastal area for several days and even weeks. This unwelcome visitor is likely to remain until the wind blows from another direction.

Another type of fog, radiation fog, is likely to form along the major portion of this coast during warm weather, provided the requirements of a clear night with little wind are at hand. Radiation fog forms in heavily wooded harbors or those surrounded by low marshy land. Early morning often finds many east coast harbors blanketed with fog, but the sun usually burns it off and full visibility is restored by midmorning. This type of fog does not extend very far to sea.

COASTAL SURFACE AND LOCAL WINDS

The prevailing surface winds blow along most of the Atlantic coast of the United States from north of west during the period of mid-fall to early spring. The velocity of these strong winter winds diminishes to an average of 7 to 16 knots in the spring when their advancement is hampered by the expanding Bermuda-Azores High. From May to September, the balmier prevailing winds blow from the southwest and south.

In the warm seasons, whenever coastal areas are in the throes of a stationary front, one hot windless day follows another. This stifling condition continues until the weather standoff is resolved between the usual west-to-east weather system progression and the Bermuda-Azores High, which stalled the front. The only relief is provided by sea and land breezes, the local weather condition that occurs on hot days along the shores of large bodies of water.

STORM WAVES AND SWELLS

Huge, breaking waves are spectacular to view from the shore, but fearsome to behold when you are out on the water. Seas can build up quickly during storms in the shallow waters near a coast and waves break with great force on shoals and on rocks even if they are several feet below the surface of the water.

In northeasterly and easterly gales, especially with an out-going current, storm waves often deny boatmen caught out in them the haven of a sheltering port along this coastline of harbors. Confused, raging storm seas, coupled with the poor visibility that often accompanies them, make it very hazardous to attempt a passage through the narrow jettied or rockbound entrances found at most of the deep-water harbors of this coast.

Near man-made entrances cut through barrier beaches to coastal harbors, the waves in these onshore blows build up to break on shoals and roll through the dredged channels churning up and shifting sandbars on the sandy bottoms as they go. This causes channels through these passageways to be relocated with such frequency that it is impossible to guarantee the accuracy of channels as shown on charts of many inlets and harbor entrances; warnings are printed on navigational charts where they apply.

Big, undulating swells resulting from storms far out in the Atlantic Ocean sometimes roll in toward the coast to break on the shore. The size of the swells in a particular area depends on the distance and intensity of the storm that caused them and the depth and contour of the ocean bottom over which they travel. Ordinarily, swells are not much of a problem for offshore boating along this coast. Handling swells in a confined area requires great attention on the part of the helmsman and should not be attempted without local knowledge, especially when they are breaking in shallow harbor entrances.

The danger of maverick ocean storms arriving unannounced on this coast is reduced now that the National Weather Service is receiving weather data from giant weather buoys located at sea. Weather Service meteorologists were convinced of the value of monster buoys after a three-year study of the performance of Giant Weather Buoy #1, a 100-ton* experimental meteorological data-reporting buoy anchored in the Gulf Stream about 110 nautical miles southeast of Norfolk, Virginia.

According to a NOAA report compiled at the end of the three-year study:

> The Atlantic Buoy (Number 1) has already made noteworthy contributions to the weather watch along the densely-populated east coast. In the three-year period in which it has been functioning, the buoy has been one of the prime sources of data which revealed the formation of more than 50 major storms of the mid-Atlantic and New England coasts.

NORTHEASTERS

When a mariner along the coasts of New England and the Middle Atlantic states hears a prediction of a storm with strong winds from the northeast or east, he knows that he is in for a seige of dirty weather. Northeasters, or nor'easters as they are called in nautical lingo, have caused considerable destruction in this region over the years.

These storms are also referred to as extratropical cyclones. The term extratropical describes the birthplace of the northeaster and differentiates it from the hurricane, which is called a tropical cyclone because of its tropical beginnings. The term cyclone, you will remember, refers to the surface winds within a low pressure area that spiral around its center; the direction of the spiral is counterclockwise in the Northern Hemisphere.

A study of the average paths of cyclones in the Northern Hemisphere reveals that those crossing the continent with the Prevailing Westerlies move from west to east. However, those that form off the southeastern or southern coasts of the United States tend to travel in a northeasterly path up the Atlantic coastline.

The low pressure areas that spawn northeasters are most likely to originate along the coasts of southern and southeastern United States from late fall through early spring. At this time, the temperature differ-

* 90 metric tons.

ence is greatest between the cold air over the land and the warmer air over the adjacent waters. As these low pressure areas move north and east, they can intensify rapidly, resulting in severe coastal storms from the Middle Atlantic states north.

In the early days of New England when fishing was the way of life and livelihood for many able-bodied men, severe northeasters made widows and orphans of many of the coastal town residents. Towns and harbors exposed to the northeast and east have sustained considerable damage during some of these storms. With a strong wind

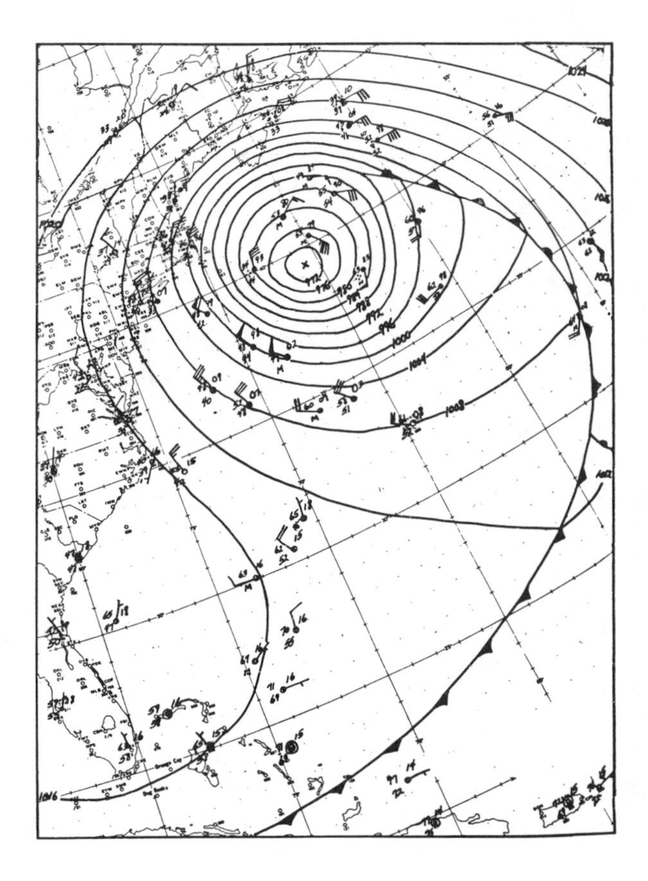

Fig. 17. A TYPICAL NORTHEASTER *This surface weather map was recorded in New England during the month of April several years ago. Published by Weather Bureau.*

Wind speed: knots

Atmospheric pressure: millibars

blowing onshore, lowland coastal regions are often flooded by the surge produced by northeasters near to or at the time of high tide. Storm surge is measured according to the excess rise in water over the predicted or normal tide. It is caused by a strong wind blowing from one direction across the open ocean for an extended period.

High winds of up to 55 or 60 knots and destructive surge can radiate for more than 450 nautical miles from the center of a northeaster. When the center of the low pressure area is far enough offshore, there may not be any precipitation near the coast although the wind and seas are still rampant and the sky may be cloudy.

Northeasters are likely to be more severe north of Cape Cod, Mass. The National Weather Service states that, "northeasters producing strong winds and high surges along the New England coast are well-developed and mature extratropical lows. The gross features of the more common northeaster consist of a single center of low pressure associated with one cold front and one warm front. These storms are usually in the initial stage of occlusion."

Note the occluded front in the upper-right-hand portion of Fig. 17, just above the junction of the cold and warm fronts.

A strong high pressure area is usually located in advance of a severe northeaster, which blocks the normal progress of the weather system. This tends to increase the rate of fall of barometric air pressure around the center of the low. Unfortunately, our weather map illustrating the northeaster does not extend far enough to the east to include this high, but its presence is inferred by the intense pressure change indicated by the tight series of isobars around the "x" marking the center of the low. As mentioned previously, the more rapid a pressure change, the greater the velocity of the surface wind will be, and this is illustrated by the numerous feathers and the scattering of pennants on the wind arrows at the locations of the weather reporting stations.

Although most of the severe northeasters occur during the cooler months of the year, one storm finished the 1967 boating season for a great many boat owners in Marblehead, Mass., before it really got started. It was the 27th of May that year when we stood on the shore struggling to remain upright against the force of the wind that was rolling huge, breaking seas down the length of Marblehead Harbor. We peered with disbelief and horror through the driving sheets of rain as one boat after another broke free of its mooring and was either dashed against the rocks lining the harbor sides or was swept against the causeway at the head of the harbor to join the ever-mounting pile of debris. Although this was the very beginning of the recreational boating season in this area, several hundred boats were already swinging at their moorings when the northeaster struck. Of these, over a hundred boats were totally destroyed, and few, if any, completely escaped damage.

Many boatmen along this coast stoutly maintain, as their fathers and mothers did before them, that all northeasters last for three days. In fact, this bit of weather know-how is so firmly fixed in seafaring lore that it seems almost presumptuous to cast doubt on its reliability. However, the weather know-how of marine forecaster Robert Lynde of the National Weather Service in Boston is in the expert class, and he tells us that there is no real basis for this assumption. It must be considered as just another old wives' tale.

HURRICANES AND NORTHEASTERS: NORTHERN NEW ENGLAND COMPARISON

Hurricanes have long had the justly deserved reputation of being among the most destructive storms in nature's repertoire. However, the surges and winds of northeasters also have potential for causing major destruction along the coastal regions north of Cape Cod, Mass. Several years ago, the Weather Bureau (predecessor to the National Weather Service) made comparative studies of the two types of storms; Fig. 18 is a portion of the conclusions reached in the study entitled, "Criteria for a Standard Project Northeaster, North of Cape Cod."

SECTIONAL CHARACTERISTICS OF THE EASTERN SEABOARD

The coastal waters off the eastern seaboard of the United States and many of its bays and sounds are considered home waters by the Townsends and the Ericsons. Between the two families, there is comparatively little of this region that is unfamiliar to us. Back in the prehistoric days of our youths, Sallie and Sam were venturing forth on Long Island Sound while Virginia and John were poling their various crafts up the creeks of Chesapeake Bay to see what was around the next bend. From these waterlogged beginnings, we branched out into other boating areas around the country, trying our hands at racing and cruising at every opportunity. Our meeting and sharing of our mutual involvement with boats and the sea occurred when Marblehead, Mass., became the home port of both families. The Ericsons still carry this hail on the stern of their auxiliary, but the Townsends now have both their home and racing/cruising auxiliary just a sea shell's throw

COMPARISON OF HURRICANES AND NORTHEASTERS IN NORTHERN NEW ENGLAND (NOAA)

Characteristics	Hurricanes	Northeasters
Climatology		
Origin	Tropics, always over water	Mostly Gulf of Mexico or South Atlantic regions, usually near coast
Season	August-October	October-April
Development	Reach greatest intensity south of New England and then diminish	Reach greatest intensity as they pass New England
Frequency	Average of 1 per 6 years	1 to 2 per year with surge equal to or greater than 2.0 feet
Track		
Direction	N to NE	N to E
Speed	Average 36 knots; range 29-48 knots	Average 22 knots; range 6-43 knots
Pressures		
Central pressure	Average 958 mb. (28.29 in.); lowest 943 mb. (27.85 in.)	Average 983 mb. (29.03 in.); lowest 957 mb. (28.26 in.)
Pressure pattern	Usually symmetrical	Usually asymmetrical
Average storm diameter	Small (400-600 nautical miles)	Large (600-1,000 nautical miles)
Winds		
Maximum speeds	80-100 knots not uncommon	70 knots is rare
Radius of maximum winds	22-66 nautical miles, well-defined	90-340 nautical miles, not well-defined, sometimes more than one
Fetch lengths	Short	Long, 300-1,400 nautical miles
Surge		
Surge heights	3.7 ft. highest observed	Up to 5.1 ft. observed
Duration of high surge and and strong wind	6-12 hours	12 hours to 3 days
Inverted barometer effect	May give important contribution to surge	Relatively unimportant contribution to surge
Topographical consideration	Cape Cod protects northern New England to some extent	Little protection afforded by Cape Cod

Fig. 18. COMPARISON OF HURRICANES AND NORTHEASTERS—Northern New England *This study was made several years ago and reprinted verbatim. Metric conversions may be found in the Appendix. Published by U.S. Department of Commerce.*

from the main channel of the Atlantic Intracoastal Waterway in Tequesta, Fla.

The surprising number of tidal waterways that lace the shores of this Atlantic coast along almost all of its entire length provide residents and visitors alike unlimited opportunities for every type of boating activity. A duck contemplating an airborne course down the Atlantic seacoast from the Canadian/Maine border to the tip of the Florida Keys can expect to fly about 1,797 nautical miles, stopping to rest here and there along the way, before he finally sideslips down to his destination at Key West. If, however, the duck's flying and swimming abilities are not up to par or he is overcome by curiosity for waterfront exploration along the tidal shoreline of the coast, including sidetrips around all the islands, he will have to webfoot a footsore, weary 24,916 nautical miles, over 23,000 nautical miles more than the airborne trip. This startling tidal measurement was published by NOAA according to their definition of this tidal shoreline as, "Shoreline of outer coast, offshore islands, sounds, bays, rivers, and creeks . . . to a point where tidal waters narrow to a width of 100 feet."*

A comparison of these two shoreline measurements, adapted from NOAA figures, for the fifteen states along the eastern seaboard are given in nautical miles as follows:

	General Coastline	*Tidal Shoreline*
Maine	198	3,022
New Hampshire	11	114
Massachusetts	167	1,320
Rhode Island	35	334
Connecticut	(no Atlantic coastline)	537
New York	110	1,608
New Jersey	113	1,557
Pennsylvania	(no Atlantic coastline)	77
Delaware	24	331
Maryland	27	2,772
Virginia	97	2,880
North Carolina	262	2,933
South Carolina	162	2,499
Georgia	87	2,037
Florida (Atlantic coast only)	504	2,895
Total nautical mileage	1,797	24,916

* 30 meters.

As the eastern coast of the continental United States spans more than 1,200 nautical miles of latitude in a direct line from north to south, certain weather patterns are more characteristic of one section of this seaboard than another. We have arbitrarily divided the coastline into four sections: Maine to New Jersey, New Jersey to Virginia, Virginia to Florida, and Florida's east coast and her Keys.

Maine to New Jersey The entire section of coastline from Maine to New Jersey is plagued with changeable weather conditions for most of the year because it lies directly in the track of moist warm fronts moving up from the southwest and blustery cold fronts pushing down from the northwest. The temperature of the air and water as well as the unleashed fury of the storms during the colder months restricts most recreational boating activities to the period of mid-May to mid-October.

Strong cold fronts are by no means confined only to winter weather. During the boating season, we have clocked gusts of 50 to 60 knots in several prefrontal squalls along this coast and have lowered sail to ride out line squalls that bore down with slashing blasts of wind and blinding sheets of driving rain.

When a boat and crew are not well equipped for storm conditions, the sensible course of action is to remain in port and wait out the passage of a cold front. (Low pressure areas moving up the coast from the south, which intensify into northeasters as they near this section, command the same attention from us as do approaching cold fronts.) In the summertime, the wait is worthwhile because once a cold front passes by, the resulting north winds of the following day or two are usually light, the air is crystal clear, and the bright sunlight dances on the surface of the clear blue water.

Summer warm fronts that become stationary when they are blocked by the Bermuda High cause hot, humid, windless days. This condition is most likely to occur south of Cape Cod. Frequent brief afternoon and evening thundershowers can be expected at this time, with the additional welcome relief from the heat often provided by afternoon sea breezes. Sea breezes are likely to vary in strength according to their locale.

The waters close to the coast of northern New England are likely to be liberally sprinkled with gaily colored lobster pot buoys bobbing on the surface, which are tethered to traps resting on the bottom of the sea. These buoys, which are also to be found in harbors, along the sides of channels, and often in the channels as well, are to be carefully avoided. Their mooring lines can foul any part of a craft that projects below the surface of the water, such as a turning propeller in particular. As lobster buoys are almost impossible to see when the visibility is poor, it is advisible to maintain an alert bow watch.

Many boatmen who cruise in the state of Maine have an alarming although justifiable tendency to become lyrical when describing the ragged rockbound coast and the crisp, clear waters of the many inlets, coves, bays, and broad rivers studded with craggy rocks and balsam-wooded islands. They seem to overlook those boating seasons when pea-soup advection fogs frequently blanket the harbors and other coastal areas of the state. The *U.S. Coast Pilot* asserts, "These fogs often set in almost without warning, and have been known to persist for three weeks almost without interruption".

Nevertheless, the Maine coast is one of the few areas along the eastern seaboard that still retains the look of untamed wilderness within reach of well-equipped harbors.

Below Portland, Maine, and on to the Cape Cod Canal in southern Massachusetts, the coastline becomes less rugged and more populated. The vistas are tamer, but still very beautiful. Conviviality is the rule at the end of a summer's day in the crowded harbors of New Hampshire and Massachusetts. This tendency to crowding is a necessary evil in these states, because anchorages are fewer and farther apart than those of their northern neighbor.

The Cape Cod Canal separates Massachusetts from its peninsula of Cape Cod, which hooks 50 nautical miles or so out into the Atlantic Ocean beyond the cut of the canal. Both sides of the peninsula's coastline are almost completely rimmed with beaches all the way to the tip of the hook. The north side of the cape forms the southern shore of the Gulf of Maine as well as Massachusetts Bay, which is part of the gulf. A branch of the cold Labrador Current sweeps down from Maine into this gulf, chilling the offshore waters and moderating the summer temperatures along the coast and the northern shore of Cape Cod. This ocean current's track through the bordering warmer waters of the ocean, however, is a chilly invitation for the formation of advection fog. This type of fog affects all of northern New England's coastal areas as well as the waters of Nantucket and Vineyard Sounds and Buzzard's Bay to the south of Cape Cod.

The Marine Weather Services Chart for the area states: "The cold water off the New England coast is responsible for heavy fog conditions in spring and summer when a warm moist southerly or southeasterly flow of air passes over the cold water. Clearing takes place when the prevailing winds become westerly or northerly."

In general, the waters to the south of Cape Cod are noted for their warmth, due to the proximity of the Gulf Stream. Here, numerous harbors line the Massachusetts south shore and the shores of the bays and sounds of Rhode Island, Connecticut, and New York. The heavy population along these shores combined with the warmer water and air temperature prevailing south of Cape Cod, make this area a beehive of boating activity, crowded to overflowing on summer weekends with large and small boats of every description.

Buzzard's Bay, just south of the cape, appears to be a natural wind funnel, situated as it is between the mainland of Massachusetts to the northwest and the Elizabeth Islands to the southeast, and the afternoon breeze can whistle up to an alarming velocity by the time it reaches the entrance to the Cape Cod Canal. Those who do the major part of their boating on Buzzard's Bay become used to strong winds and find other areas tame by comparison.

According to the *Marine Weather Services Charts:* "Long Island Sound weather during the boating season is known for light winds and late arriving sea breezes. However, prolonged northeasterly or southwesterly winds are much stronger over the water and produce uncomfortable seas on the down wind end of the sound."

On light-air days, sea breezes in the narrow western end of the sound are elevated by the high buildings on Long Island so that they blow above the water, often missing it entirely.

At the New York end of Long Island Sound, southbound recreational boaters headed for New York Harbor join a stream of commercial traffic traveling the East River. If they have done their homework, they will time their arrival at the swirling eddies of Hell Gate in the East River at slack water. An arrival time at high-tide slack water allows them to benefit from the current on the way to Hell Gate from Long Island Sound as well as after leaving it.

A trip on the East River and the river's-eye view of the great metropolis of New York City is an unforgettable experience. It is a friendly time, everyone waves, those on other boats and those ashore. It almost seems that the wind lays in wait for the happily distracted helmsman to blast out in gusts across the water, channeled through the apartment-lined streets that end at the river's edge.

On past the United Nations and the Battery, and out into the Upper Bay of New York Harbor, keeping an alert eye peeled to avoid colliding with the ferries, tugboats, and barges of every description scuttling about like so many water bugs on a pond, and then there she is, the Statue of Liberty. The lady is serenely unaffected by the confusing array of work boats, freighters, liners, and airplanes around and above her, and the floating debris at her feet.

Dwarfed in comparison with the great ocean-going ships, you may feel as apprehensive as we do about crossing the main channels of the Lower Bay to the safety of the New Jersey shore. Those huge ships require a mile or so to stop and they certainly do not turn on a dime.

New Jersey to Virginia From New York Harbor to the start of the New Jersey Intracoastal Waterway, all southbound boats must make an ocean passage of approximately 20 nautical miles along the New Jersey coast from the tip of Sandy Hook to Manasquan Inlet. Some persons elect to continue the ocean passage all the way to Cape May and often this is the first blue-water experience for many yachtsmen from Long Island Sound. Miles and miles of sandy beach with occasional brightly painted water towers to identify a coastal town dominate one side of the scene, in sharp contrast to the stark horizon on the other side.

It is important to check weather reports carefully before embarking on this outside passage. A harbor of refuge will be denied you along most of this New Jersey coastline during an onshore blow. Conditions likely to be encountered along this coast are described in the *U.S. Coast Pilot* as follows:

> Prevailing winds at most stations are from northwest during the cooler months, October through March, and from the southwest, May through September. The average wind speeds during the warmer months are generally lower than during the colder seasons . . . . Highest average speeds occur in March and lowest in August.
>
> In the winter, the winds over the open ocean are slightly stronger than those over land. Little difference is apparent in summer. In the warmer seasons, a daily shift in wind direction occurs . . . . During the warmer part of the day winds blow onshore, and during the cooler part, offshore.
>
> Gales (force 8* or higher) are reported in about ten percent of ships' observations in winter. Gales are generally from the westerly quadrant. Summer gales are rare, but may be encountered during tropical cyclones or local thunderstorms . . . . Depths of 7 fathoms* are found as far as 13 miles from shore . . . . Gales from northeast to southeast cause heavy breakers on the beaches and outlying shoals; the seas break in 4 to 5 fathoms of water, and shoals of that depth or less usually are marked during easterly

* Beaufort Scale of Winds, Fig. 6.
* 1 fathom equals 6 feet or approximately 2 meters.

gales. The bars across the inlets are then impassable and are defined by breakers even in comparatively smooth water with a light swell.

From New Jersey to Virginia, the strength of cold fronts is moderated by the Appalachian Mountains before they reach the coastal area. During the summer, the influence of the Bermuda High provides light winds from the south and southwest. the warm humid weather during most of this season results in frequent afternoon and evening thundershowers near the coast. Fog in this area will usually dissipate by afternoon.

Many cruising skippers elect to continue on to the Absecon Inlet at Atlantic City rather than cope with the shallow depths and shifting sandbars of the Manasquan to Atlantic City section of the New Jersey Intracoastal. Friends of the Townsends, cruising in a 40-foot powerboat in one section of this waterway, said that even after reading about the area where Route 88 crosses the canal at Point Pleasant just as the waterway leaves the Manasquan River, they were unprepared for the severe turbulence they encountered there. They compared the experience to running the rapids and said they were told of many boats damaged and even lost each year at this dangerous place.

The chief of the Operation Division of the Corps of Engineers in this area suggested that "the safest time to navigate the canal is when the boater can plan the trip to arrive at the Route 88 bridge near the time of slack current condition, which generally occurs three hours after the high or low water state of the tide at the north end of the canal." In addition, he mentioned that the wind conditions are extremely important factors in relation to the tide levels and resulting current conditions, and should be considered when picking a time for approaching and passing through the bridge.

The charted depth of the New Jersey waterway is somewhat deeper from Atlantic City to Cape May Harbor than it is to the north. Passage along here for auxiliaries is limited to those requiring no more than 11-meter (35-foot) clearance because of the height of the two fixed bridges.

Southbound yachtsmen sometimes elect to take the outside passage from Cape May, N.J., to Norfolk, Va., via Chesapeake Bay along the ocean shore of the Delmarva Peninsula (so-called because the peninsula is comprised of the state of Delaware and portions of the states of Maryland and Virginia). Several yachtsmen of our acquaintance have experienced unexpectedly severe weather during this passage and have radioed the Coast Guard for advice about entering harbors of refuge. In each of these cases, the Coast Guard sent a boat out to lead them into port because of the shoal inlets.

The existing 125 nautical miles of waterway behind the Delmarva barrier islands is very shallow at present, although there is a proposal by the Army Corps of Engineers to dredge a channel along it to a depth of 2 meters (6 feet). Now, this waterway can accommodate only very shallow draft boats of the rowboat, canoe, or small outboard motor class.

The inland route to Norfolk, Va., from Cape May, N.J., goes up Delaware Bay, through the Chesapeake and Delaware Canal, down almost to the mouth of Chesapeake Bay, and through Hampton Roads past Norfolk to the northern terminus of the Atlantic Intracoastal Waterway.

Delaware Bay has a very poor reputation in long-distance cruising circles because of the short steep seas that build up very quickly whenever there is a strong wind blowing against the current. The ride up or down the bay, however, is exhilarating when the wind and current are with you. This is a pleasant area for residents and vacationers, but for long-distance cruisers there are few harbors.

The 16-nautical-mile-long Chesapeake and Delaware Canal is the gateway to the beautiful, sprawling Chesapeake Bay. The bay stretches for nearly 170 nautical miles south from the Susquehana River to the Atlantic Ocean. The shores along its entire length are designated the Eastern Shore on the Delmarva Peninsula side and the Western Shore on the Washington, D.C., side. More than thirty rivers empty into the bay and there are also countless inlets, harbors, creeks, and coves along the shores behind peninsulas, islands, islets, and sand spits.

The boating season starts in early April and continues until the end of October or even longer. In the summertime, the breezes are from the south or southwest and generally vary from light to flat calm. Although it is often chilly early and late in the boating season, sailors especially maintain these are the best months to be on Chesapeake Bay. Since this bay is a mecca for all types of boating, commercial as well as recreational, the less crowded conditions that prevail in the spring and fall are particularly attractive.

Chesapeake Bay watermen have a great respect for the force of the winds brought by sudden thundersqualls. These occur more frequently in the summer and are likely to arrive in late afternoon with much sound, fury, driving rain, and even hail for a short time, and then pass on. The *Marine Weather Services Chart* for Chesapeake Bay notes: "In Chesapeake Bay the strongest winds associated with thunderstorms generally occur during the afternoon and early evening hours. Keep informed of thunderstorm activity by the static on your AM radio, surveillance of the western horizon from the southwest through north and radar reports on NOAA Weather Radio."

As a small child sailing on the Eastern Shore in her converted rowboat, Virginia's only boating companions were the fishermen and crabbers. They taught her all they felt she needed to know about the summer weather on the bay. "When yuh see them black clouds a-makin' up on the Western Shore, yuh got a half hour to turn tail and head for shore," her sailing mentor Capt'n Jake Highland cautioned, repositioning his cud of tobacco. "Now if'n yuh were sailin' on t'other shore, yuh'd only git 'bout 10 minutes!"

Virginia to Florida Boating enthusiasts in the southeastern United States can pursue their sport during almost all of the year if they maintain a careful weather watch and on occasion a die-hard attitude. A great deal of the boating takes place inside the barrier beaches and islands that line this section of the coast. The many passages to the sea, however, lure fishing, racing, and cruising devotees out into the open ocean. Fishing in the Gulf Stream off this section of coast is so important, both commercially and for pleasure, that weather reports often include offshore weather conditions.

The Blue Ridge section of the Appalachian Mountains stretches from Virginia to northern Georgia. Since this range has hundreds of peaks over 1500 meters (5,000 feet) in height, it serves as a partial obstruction to many of the winter storms headed for the coast. The storms that do manage to push through, however, seem to have been slowed only slightly, bringing as they do cold, often freezing temperatures, strong gusty north to northwest winds, and sometimes even snow. On the average, the winter winds are only 8 to 15 knots along this section of coast and the temperatures are moderate. As the Bermuda-Azores High is far out to sea during this season, changeable air masses often follow each other in close succession without obstruction, some of which are accompanied by fog.

The seasonal movement of the Bermuda-Azores High brings it nearer the coast in the spring, and with it come fair-weather southwest winds. Rain and warm, sometimes hot, humid weather also occurs during this season. Radiation fog often forms during the night along the coast, which no doubt inspired the legendary ghost stories of The Gullah Country, that part of the coast from north of Charleston, S.C., to

Cumberland Island in Georgia. This area abounds with tales of haunted spirits, which are said to roam the desolate lowlands especially in the wispy fogs of early morning.

In the summertime along this section of the coast, the winds blow more from the south and southwest and are usually very light, averaging 6 to 10 knots. The weather can be hot and humid for weeks at a time accompanied by localized rain and thundershower activity.

The hurricane season officially begins June 1, although these monster storms have been known to occur as early as May or as late as November. If you find yourself and your craft within the possible track of an approaching hurricane, the following suggested procedure for mooring in this area, as quoted from the *U.S. Coast Pilot*, should be of value:

> On receiving advisory notice of a tropical disturbance, small boats should seek shelter in a small winding stream whose banks are lined with trees, preferably cedar or mangrove. Moor with bow and stern lines fastened to the lower branches; if possible snub up with good chafing gear. The knees of the trees will act as fenders and the branches, having more give than the trunks, will ease the shocks of heavy gusts. If the banks are lined only with small trees or large shrubs, use clumps of them within each hawser loop. Keep clear of any tall pines as they generally have shallow roots and are more apt to be blown down.

After leaving Chesapeake Bay, the southbound cruiser will be in Virginia for just a short distance before crossing the North Carolina state line. Even experienced blue-water yachtsmen, who have chosen an ocean passage for most of their southbound trip, usually abandon this route just north of Cape Henry for a short trip on Chesapeake Bay in order to pick up the Atlantic Intracoastal Waterway. They continue south on the waterway, at least from Norfolk to Morehead City, N.C., situated just south of Cape Lookout, before resuming their ocean passage.

This inland detour avoids the dangerous waters off Cape Hatteras, on one of the North Carolina barrier islands. These waters are known by the gruesome title, "The Graveyard of the Atlantic." Shoals here extend as far as seven nautical miles out to sea. Unpredictable winds are said to cause the buildup of six-meter (20-foot) seas in minutes causing sand bars to shift and creating dangerous rip tides and variable currents. A chart is available at the nearby Wright Brothers' National Memorial Visitors' Center showing the location of more than

500 wrecks along the 175 nautical miles of coastline between Cape Henry to the north and Cape Lookout to the south of Cape Hatteras.

It is an interesting historical sidelight that Wilbur and Orville Wright selected their lonely windswept site for glider experiments and preflight tests after being informed by the Department of Agriculture in 1900 that the winds blow along these barrier islands from north to northwest in September and October. The results of their experiments startled and changed the world when they achieved man's first successful flight in a power-driven heavier-than-air machine on December 17, 1903.

The *U.S. Coast Pilot* advises those who throw caution to the wind and continue the outside passage past Cape Henry and around Cape Hatteras as follows:

> The entrance to every harbor on this stretch of coast is more or less obstructed by a shifting sand bar over which the channel depth is changeable. The entrance channels of the larger and more important harbors have been improved by dredging; in some cases jetties have been built from both sides of the entrance.
>
> The buoys on many of the bars are not charted because they are moved from time to time to indicate the changing channel. They are liable to be dragged out of position and cannot always be replaced immediately, so a stranger must use the greatest caution. . . .
>
> Extreme wind velocities are a hazard in any month. Though winds greater than 34 knots are comparatively infrequent, they have been recorded at all stations in this stretch of coast at almost any time of year.

The Atlantic Intracoastal Waterway from Chesapeake Bay southward is a study in contrasts, offering a gamut of experiences from relaxed cruising surrounded by serene beauty to challenging problems in seamanship and navigation. Small craft navigational charts for each part of the waterway to be traveled are essential. Included in the information on them are consecutive mileage figures starting with zero miles at Norfolk. The waterway traverses four states north of Florida and its approximate nautical mileage within each of these states is as follows: 29 miles in Virginia, 297 miles in North Carolina, 204 miles in South Carolina, and 120 miles in Georgia.

Waterway travel is not always a carefree, relaxed cruise in confined, safe waters with no weather problems to be encountered along the way, as many believe it to be. There are many protected canals along the Intracoastal, it is true, but these canals are connections for natural bodies of water such as rivers, lakes, sounds, and bays, many of which have considerable expanses of open water. Winds sweeping across the barrier dunes of North Carolina can cause very rough water in the wide, shallow sounds adjacent to the waterway route. The wind also increases in speed as it funnels into a river or through a harbor entrance, and wicked seas occur when a strong wind blows in the opposite direction to a powerful current.

In the more open spaces, crosscurrents and winds can insidiously sweep a boat out of the prescribed channel. A careful skipper, particularly aboard a slow-moving boat, frequently looks behind to be sure the boat is still within the channel, adjusting as necessary to correct for any sideways slippage.

In general, only large commercial tugs travel at night on the waterway. Although the captains are usually familiar with the route, the vessels are equipped with powerful searchlights for scanning the banks to determine the distance off and for picking out markers. For most waterway travelers, however, sundown is a major factor in deciding where to spend the night, because there are many stretches where marinas are scarce and protected anchorages are few and far between. In their many trips on waterways, the Townsends have been known to use the following inexact, although still helpful, method for determining the time of sundown.

This method is based on the fact that the sun nears the horizon at a rate of 15 degrees per hour. Hold your hand out in front of you with your thumb on the bottom parallel to the water and pointing at the horizon. The rest of your hand is held flat with your fingertips pointing at the sun. You can estimate the number of hours to sunset from the angle formed between your thumb and forefinger by eye or by using your compass. With the latter method, keep your hand in this position and turn it so that you can look at it in relation to your compass. Think of the apex as being at the center of the compass and estimate the approximate degrees. If the angle appears to be 45°, for example, the sun will set in about three hours. (Care must be taken not to look directly at the bright sun.)

Northern yachtsmen planning for the first time to take their boats to warmer climes in the fall are likely to think they will see the end of cold and stormy weather once they enter the Intracoastal Waterway at Norfolk. The weather will moderate as they progress southward, but the cold, hard facts of the matter are that the impact of late fall and winter storms is felt all along the eastern seaboard.

While stormy weather is not always on the waterway agenda in the late fall, the following November entries from several of the Townsend's logbooks should alert boaters to the possibility of encountering similar conditions:

> *Norfolk, Virginia.* Excitement at marina as we tied up because two tornadoes accompanied a strong cold front's arrival this afternoon.
>
> *Morehead City, N.C.* Laid over today as below freezing; windy, strong cold front came through yesterday.
>
> *Isle of Hope, Georgia.* Forecast for winds northeast 15 to 20. Decided to look at St. Andrews Sound to see if we can make it. The current is flooding this morning, and wind and current, although both against us going to outer buoy, are from the same direction. 0935—rounded outer buoy. It is blowing 30 to 35 and the water is rough. Wish we had been able to take the alternate route today, but this boat draws too much water (6 feet*).
>
> *Jacksonville Beach, Florida.* Thanksgiving. Dismal. Wind NE 15–25. Temperature forecast 33°F (1°C). 1445—arrived Marineland in snowmobile suit, furry hat, boots, warm gloves, and a cold nose.

Florida's East Coast and Her Keys The peninsular state of Florida reaches into the most southerly latitude of the continental United States. It is bounded on the east coast by the Atlantic Ocean and on the west by the Gulf of Mexico. The Florida Keys are a long, low chain of islands that stretch from the tip of Florida's mainland toward the southwest for a distance of about 150 nautical miles.

Most of the state of Florida is south of lat. 30°N and therefore juts into the terrestrial wind belt of the Northeast Trades. Along the east coast, the trades are augmented almost daily by a sea breeze because of the heat of the land in this semitropical state. This cools the coast, but presents a potentially hazardous boating situation at the mouths of inlets and harbors when the wind is strong and blowing against an ebb current even in pleasant weather. Most entrances to the sea are vulnerable to steep breaking seas throughout the year as they are either confined by breakwaters or crossed by ridges of sand not far below the surface of the water.

As in many other regions, winter weather here can vary greatly from year to year despite the tropical trend of the weather patterns. As one Gold Coast resident put it after a couple of unusually cold, rainy, windy winter months, "And to think, last year we did not have any

---

* 1.8 meters.

winter at all!" From our observations, during some winters cold fronts go through this region on an average of about twice a week. Many have no precipitation and are barely noticeable except for a wind shift toward the north. Some of these cold fronts, however, are accompanied by two to three days of steady rain while others are ushered in with strong winds and driving rain of short duration.

Barometric readings are not too dependable for alerting boaters to the severity of an approaching storm and the probable strength of the wind. The passage of low and high pressure systems barely registers on barograph charts. Most boating enthusiasts rely on nature's announcement of an approaching cold front—when the wind moves around into the south from its prevailing easterly to southeasterly direction, boaters beware! It almost always continues veering to blow from the west and finally from the northern quadrant. The stronger the wind blows from the south, the more likely severe weather will be associated with the front.

Many persons who cruise or fish off Florida's east coast contend that it often is not only the lows of cold fronts that cause strong winds during the winter, it is the highs as well. When a large high pressure area follows a low across the lower half of the continent, it often passes just to the north of Florida's border and becomes stationary just off the coast. The outward clockwise circulation of air from the high plus the Northeasterly Trades and daily sea breezes produce blustery northeast winds down much of this coast.

The incidence of fog decreases considerably the farther south you go. For example, the *U.S. Coast Pilot* reports an annual average of thirty-three days of heavy fog at Jacksonville, eight days at West Palm Beach, and only one day at Key West. Fog forms more frequently in the northern sections of the state because of the likelihood of a sharp drop in temperature, and sometimes harbor entrances are blanketed for hours. When fog occurs in the middle section of this Florida coastline, it is usually of short duration. When fog creeps into the Keys, it is headline news; radio alert warnings are broadcast both for boaters and automobile drivers.

You can rely on hot and humid weather for days on end during the summertime, which is alleviated slightly by the on-shore sea breeze that is likely to blow from mid-morning until late evening. At least one brief tropical thunderstorm or cloudburst can be expected every day, and sometimes several are experienced in the same day interspersed with clearing skies and high humidity. During August and September, torrential and long-lasting rains can be expected whenever a tropical disturbance is near the area.

Several large boating communities in Florida have formal plans for moving boats away from the coast and into canals and rivers when hurricanes are expected. There are some places along the Florida portion of the Atlantic Intracoastal Waterway that are closed to regular boating traffic when a hurricane is forecast so they can be used solely as anchorages during the storm. Persons mooring their boats in creeks and rivers prior to a hurricane should be very careful not to obstruct the passage of other boats heading further upstream.

Those contemplating an outside or ocean passage along the Florida coast should spend some time researching the Gulf Stream prior to making the trip. As you know from the information on ocean currents discussed earlier in this chapter, the Gulf Stream sets to the north, and even a mild north wind develops rough water in the stream, a condition that can be dangerous for small craft. The Gulf Stream hugs the shore from the Florida Straits to Jupiter Inlet, bringing beautiful warm, clear turquoise water to the coastal shallows, and from there it moves gradually away from the coast. The Gulf Stream moderates the summer temperatures as well as the winter ones and keeps the mercury from climbing over the 38°C (100°F) mark as far south as Key West.

The Intracoastal Waterway from Cumberland Sound at the northern border of the state to Miami and the start of the Keys is about 325 nautical miles long. One of the greatest problems encountered while traveling Florida's section of the waterway is the great number of bridges that cross it. This is of particular concern aboard boats with flying bridges or masts since most of the bridges will have to be opened for them. To accommodate the requirements of both waterway and land traffic, many of the bridges open only twice an hour and others are closed for an hour or more in the morning as well as for the same length of time in the afternoon. *Movable Bridges,** a leaflet published for the convenience of the waterway traveler, gives the opening regulations for these Florida bridges. It also includes information on hazardous currents that might be encountered at the bridges.

Inland waters become choppy to rough when the velocity of the wind is high. The wide expanses of Mosquito Lagoon and Indian River, which comprise about one-third of the eastern Florida section of the Intracoastal, are particularly susceptible to wind-churned waters because they are so shallow.

The Intracoastal Waterway route to the Florida Keys starts in

---

* Distributed by Florida Inland Navigation District, 2725 Avenue E, Riviera Beach, Fla. 33404.

Biscayne Bay at Virginia Key and runs along the Gulf of Mexico side of the Keys. Most of the Keys are low coral and sand formations and are covered with spiderlike, aerial-rooted mangrove trees. Although the waterway route to Marathon on the lower Keys is picturesque, sections of it are only 1.5 meters (5 feet) deep and much of it between Marathon and Key West is really nonexistent.

Higher winds are commonly registered at the Key West weather station than at other Florida stations. The winds in the Keys cause shifting of sand bars, and, therefore, the shoal areas are uncharted. Since this waterway lies to the north of the Keys, the route is open to the strong north winds that sometimes accompany a winter cold front. At this time, the Keys become a dreaded lee shore where a vessel can be blown toward land to founder on the jagged coral islands.

When the shallow waters are churned up by wind, current, or even the suction of powerful propellers, they take on a milky appearance. At most times, however, boaters can expect to see sparkling clear, clear water interspersed here and there with shades of very pale green, which warns of sandy shoals, and dark mustard, which warns of coral reefs. The clarity of the water—it is possible to see the bottom at depths of as much as 3 to 5 meters (10 to 15 feet)—is startling.

The preferred route to Key West is Hawk Channel, which runs along the southern side of the Keys and has a controlling depth of 2.7 meters (9 feet). This channel begins at Cape Florida at the southern tip of Key Biscayne. It runs between the shoal, rocky Florida reefs and the Keys. Because of the predominance of east to southeast trade winds in the region, there is almost always some surge or swells in Hawk Channel. Included in the discussion of Hawk Channel in the *U.S. Coast Pilot* is the following information:

> Light-draft vessels bound southward and westward, may use this channel with great advantage, avoiding entirely the adverse current of the Gulf Stream and finding comparatively smooth water in all winds, except when passing the large openings between the reefs in southerly winds. These openings are principally between Alligator Reef Light and American Shoal Light . . . .
>
> Possible crosscurrents should be guarded against especially in the vicinity of the openings between the Keys.

These islands are connected to the Florida mainland by the Overseas Highway, the southern extension of Route 1, which runs from Key Largo to Key West. The number of passages available to many yachtsmen who wish to go between the Intracoastal Waterway and Hawk Channel are limited. Only two routes spanned with opening bridges are recommended on the charts: Channel Five, which is 70 nautical miles from Miami, and Moser Channel, which is 13 nautical miles farther on.

The eastern edges of the Florida Keys are washed by the Straits of Florida where the confluence of several currents forms the powerful Gulf Stream. As would be expected, therefore, the current can run very rapidly between the Keys. The drift of the current in some of these passages has been reported to exceed 10 knots when a strong wind has been blowing in the same direction as the set of the current during periods of especially high and low tides. In addition, there is the danger of overfalls or short breaking waves in these passages when the current is particularly strong. These waves are most likely to develop when a strong wind encounters an opposing current and small boats have been known to be swamped while attempting a passage.

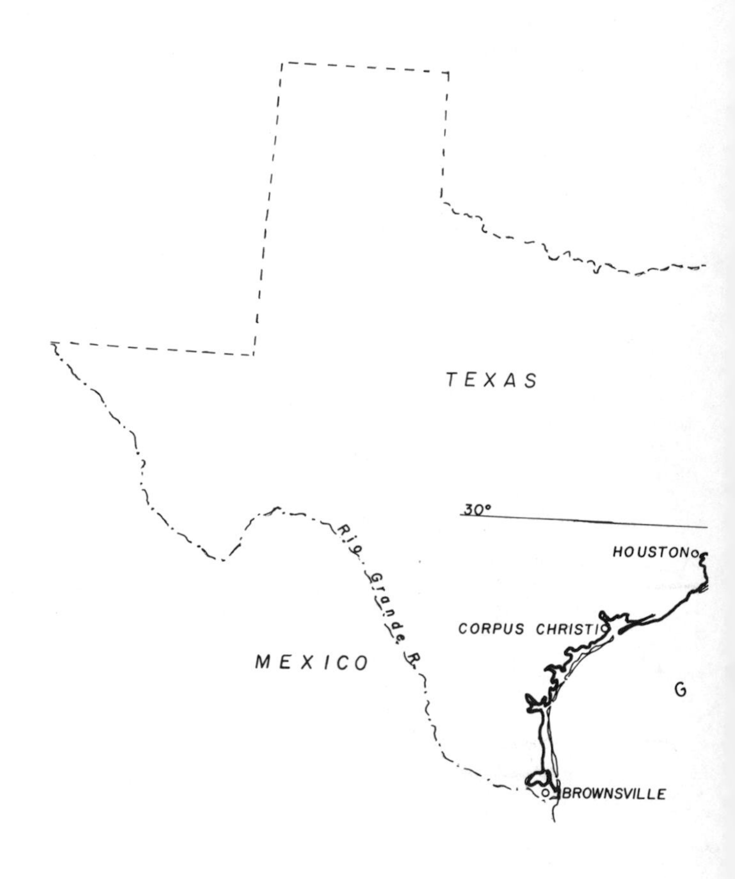

Fig. 19. THE GULF COAST—Including the Okeechobee Waterway and the Mississippi River

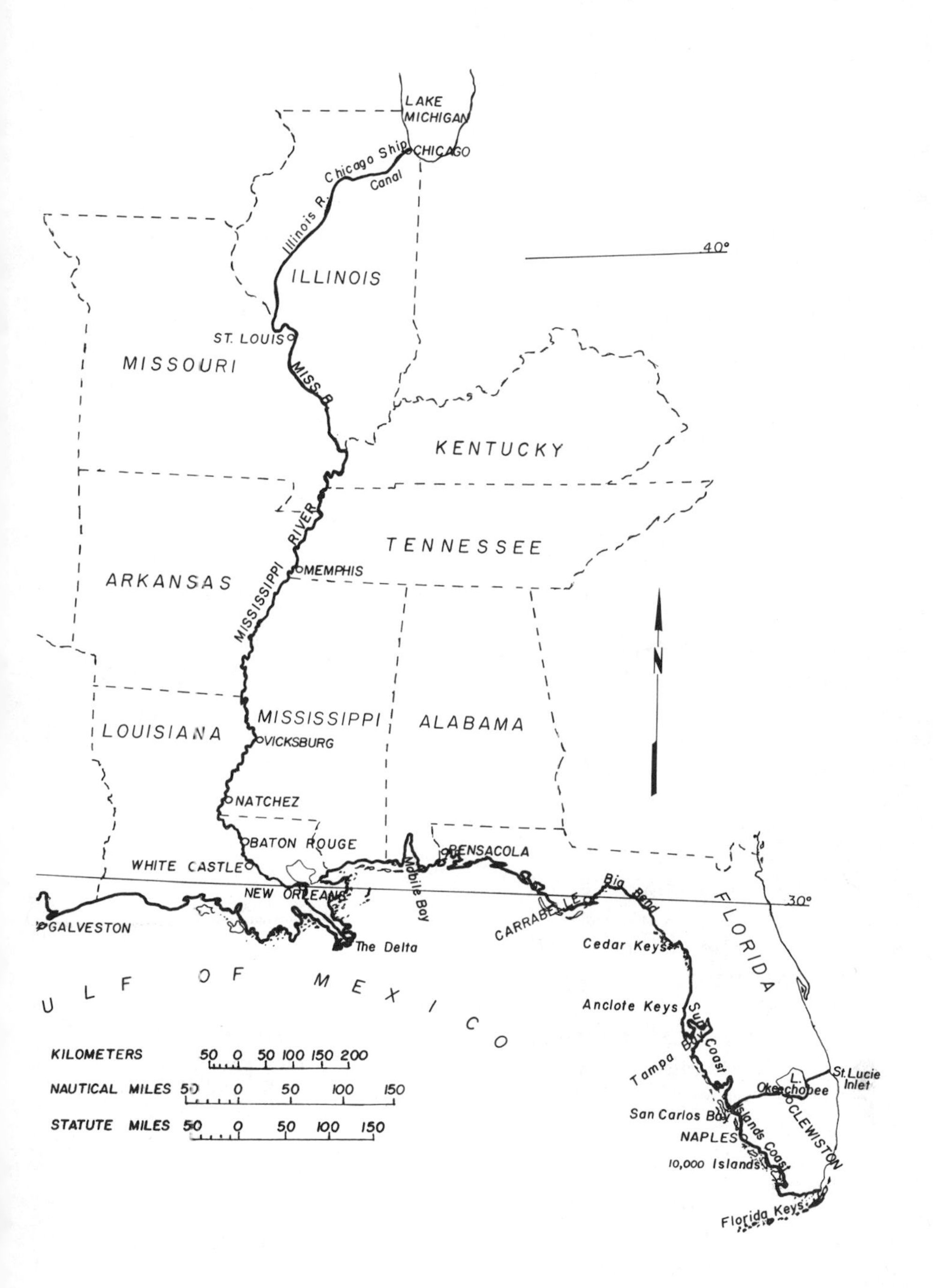
LAKE MICHIGAN
CHICAGO
Chicago Ship Canal
Illinois R.
ILLINOIS
40°
ST. LOUIS
MISSOURI
MISS. R.
KENTUCKY
TENNESSEE
MISSISSIPPI RIVER
MEMPHIS
ARKANSAS
N
MISSISSIPPI
ALABAMA
LOUISIANA
VICKSBURG
NATCHEZ
BATON ROUGE
PENSACOLA
WHITE CASTLE
Mobile Bay
NEW ORLEANS
Big Bend
CARRABELLE
30°
FLORIDA
GALVESTON
The Delta
Cedar Keys
GULF OF MEXICO
Anclote Keys
Sun Coast
Tampa Bay
KILOMETERS 50 0 50 100 150 200
NAUTICAL MILES 50 0 50 100 150
STATUTE MILES 50 0 50 100 150
L. Okechobee
St. Lucie Inlet
San Carlos Bay
CLEWISTON
NAPLES
Islands Coast
10,000 Islands
Florida Keys

CHAPTER 10

Gulf Coast Weather

The shores of the United States lining the Gulf of Mexico are steeped in the romance of waterborne conquerors and buccaneers who used them as a gateway to territorial expansion, commerce, or just plain personal greed. Residents with roots in the earliest settlements reflect Spanish and French heritages, unlike the largely English and northern-European heritages of the majority of early Atlantic coast settlers.

Five southern states—Florida, Alabama, Mississippi, Louisiana, and Texas—form a crescent around the northern shores of the Gulf of Mexico. Many major rivers carrying runoff from 40 percent of the continental United States flow through these states to empty into the Gulf of Mexico.

The eastern end of this U.S. coastline starts at the tip of the Florida Keys. It arcs northward to the bend in the Florida coast at the state's panhandle and then continues westward to Texas on or near lat. 30°N. The coastline dips to the south again along the Texas shores to the Rio Grande River and the Mexican border.

Most of this coastal region is low-lying and consists either of sandy barrier islands or junglelike marshland. Since the climate is semitropical, recreational boating can be enjoyed practically year around.

GULF COAST TERMS

It soon becomes apparent to a stranger cruising in these waters that the residents have a descriptive language of their own for terms commonly used elsewhere in the country. This no doubt stems from the multilingual background of the region. The residents in certain of these coastal areas are noted for their fierce independence, and waterborne visitors quickly find it practical to adopt the local terminology. For example, an inlet from the Gulf of Mexico into a waterway is called a *pass*, a sluggish watercourse draining from swamplands or rivers into the gulf is a *bayou*, and an island in Florida's gulf waters is a *key*. An unusually strong, fast-moving winter storm blasting cold northerly wind across the inland plains and into the Gulf of Mexico is referred to as a *norther*.

The *U.S. Coast Pilot* describes some of the local terminology that will be encountered on the Mississippi River in that section of their publication as follows:

> The Mississippi River empties into the north central part of the Gulf of Mexico through a number of mouths or passes which, taken together, form the *delta* of the river . . . . *Flocculation*, locally known as slush, is a living mass of jellied material or muck deposited in the lower part of the Mississippi, during the low stages of the river . . . . *Sand waves*, the material brought down during high stages . . . is of a sandy nature . . . . *Mud lumps* are small oval-shaped mounds or islands no more than 8-feet high, which are peculiar to the Mississippi River delta.

CURRENTS

The Gulf of Mexico is fed by an arm of the warm Equatorial Ocean Currents that sweeps in from the Caribbean Sea. The northern thrust of this current splits approximately near the Mississippi River delta with one branch flowing to the west and the other to the east. The drift of the current is generally weak in the gulf itself, but can be very strong along this northern coast especially at the mouths of the rivers.

Action from the current can change bottom contours and, near this coast, bars and shoals shift in channels and passes. An understatement of this ever-changing condition was made by the Ericsons' friends Roger and Brita Wooleyhan, who said, "A chart over six months old is out of date."

River flow, especially in the high water stages of springtime, strongly influences the set and drift of current near the mouths of rivers and raises the level of the water in nearby bays and waterways.

The average high and low tidal range along most of this gulf coast is less than 1 meter (about 2 feet) with only a few of the stations predicting 1 to 1.2 meters (3 to 4 feet) of periodic tide level difference.

Information published in the *Tide Tables* and *Tidal Current Tables* for the gulf coast may not always be completely accurate because of the effect of other forces and must be relied on as predictions, which they are, and not as gospel.

WIND

Whenever the wind is blowing strongly along the length of a waterway, it can substantially increase a current or even create one. It can cause considerable increase in the water depth of bays and waterways whenever it blows gulf water through the passes, and, conversely, the wind can cause a decrease in the water level when it blows toward the gulf. As you know from the description of wind-driven waters in the Great Lakes chapter, a sustained wind blowing for some time in the same direction can lower the water depth on one side of an enclosed or semienclosed body of water while raising it on the other. The possibility of this occurring must be considered when boating on the shallow waters of some of the bays and lakes of the Gulf Waterway. Hurricane winds have caused dangerous storm surges along this coast that have raised the water level by as much as 3 to 8 meters (10 to 25 feet), according to the National Weather Service.

WINTER NORTHERS

Winters on the Gulf of Mexico are generally mild. The rigorous, fast-moving cold fronts that plague the northern states during this time of the year tend to slow down and moderate by the time they reach the southland. A southerly wind blowing over the gulf usually exerts a benign and calming influence on these cold fronts, although there are times when winter weather kicks up its heels even here.

The *U.S. Coast Pilot* advises that some 30 to 40 polar air masses penetrate the region each winter and 15 to 20 of these cause storms with strong northerly winds. When seamen call a storm a norther, they usually mean one in which the wind blows at least 20 knots. Winds of 25 to 50 knots have been clocked in the gulf region during severe northers. One to six severe northers are likely to occur each year and residents expect the duration of each storm to vary from one to four days.

One old-timer in Texas explained the violence of northers by commenting, "Thars nothin' b'tween us'n the no'th pole excepten a barbed wire fence!" These storms are spectacular, and the Townsends observed many of them during the twelve years they lived in Texas. In their experience, a day when a norther came in usually started out mild with nature's clue to its approach visible over the northern horizon: several hours before the arrival of a norther, the entire northern sky is filled with a blue-black cloud bank, which accounts for the local descriptive term of blue norther.

These storms swoop down from the north with unleashed fury bringing sudden whiplashing winds and savage blasts of driving rain. The air temperature drops with extraordinary speed, although along the coast the gulf moderates this sudden change in air temperature from the extremes experienced inland.

According to Billy J. Crouch, acting meteorologist in charge of the National Weather Service forecast office in New Orleans: "The primary effect in shallow coastal waters and large coastal lakes is a rapid buildup of seas with a very short wave length; a very rough chop. In swamps and bayous along the coast, the waters run out very rapidly and boaters can become stranded for several days in shallow water."

A large barometric air pressure change usually cannot be expected in this area even when a winter storm is on the way. Any small change in barometric pressure during its daily fluctuation of two rises at 10 A.M. and 10 P.M. and two dips at 4 A.M. and 4 P.M. is probably a signal of the approach of inclement weather.

FOG

Whenever cold air from the north invades this region, bringing chilling rains, the surface temperature of the gulf waters is lowered. As you would expect, when warm, moist, southerly breezes then blow over the cooled water, advection fog will form. A dense fog can develop near the shore or be blown toward this coast to blanket harbors and waterways until the wind shifts to the north and blows it away.

When the waters of a river become chilled in the same fashion, river fog forms over the river as it courses between warmer land surfaces. Viewed from a low-flying airplane, this fog frequently marks a river's path as it drains into bays or harbors, and sometimes during high water on the Mississippi River cooler river waters extend 35 to 40 nautical miles into the coastal waters of the gulf itself. This area is particularly susceptible to advection fog primarily in January, February, and March. Duration can be several days with shipping at a standstill trying to enter or leave the river.

A river swollen with cold winter rains or thawing northern snows in the spring is likely to jet its way far out into the gulf before it mixes with these waters and loses its identity. Advection fog usually develops when humid surface air above the warm waters of the gulf is blown over the invading cold river water.

The numerous oil and gas rigs located in the gulf are a navigational hazard during times of lowered visibility although those near navigable channels are required to be equipped with sound apparatus and lights.

SUMMER WEATHER

From May to November, when the Bermuda-Azores High is at its peak strength, it exerts a major influence on gulf weather. This influential high pressure area is generally located near the Atlantic coast during this period in the vicinity of lat. 30°N, the same latitude as that of the longest stretch of United States coastline on the Gulf of Mexico. The far-reaching effect of the high's spiraling clockwise circulation of air deflects many eastward-moving weather systems away from this region, with the result that gulf coast weather is likely to be hot and humid from spring through fall.

Prevailing summer winds over the gulf blow from the south to southeast. This warm, moist, southerly air moves across the gulf to produce daily showers over most of the coastal areas. The entire length of this coast, with the exception of the southernmost coastal section of Texas, experiences the highest incidence of thunderstorms of any United States coastline, with the likelihood of 80 to 90 thunderstorm days each year. This scattered thunderstorm activity usually occurs in the afternoon.

Gulf coast weather is also affected by a distant belt of low pressure called the Equatorial Trough. Located near the equator, this low pressure trough is better known to mariners as the doldrums, famed for its stiffling oppressive calms and light but fickle winds. According to information sent to us by NOAA: "The segment of this belt lying to the south of the Gulf of Mexico migrates from 8°N to 10°N latitude, reaching to about 12°N latitude in July. This migration influences the seasonal march of cloudiness and rainfall and the formation of tropical storms."

TROPICAL STORMS

The National Weather Service's National Hurricane Center in Miami keeps a careful eye on weather satellite photographs for any evidence of the development of whirling wind storms over tropical waters. Whenever one is sighted, they subject it to constant surveillance and issue frequent reports concerning its size and path to be broadcast over radio and television and to be published in all the newspapers in the region. Also, during hurricane season, they are very likely to include the good news in their weather releases whenever conditions are quiet and there is no threat of a tropical disturbance on the current weather scene.

Hurricanes originating in the lower Caribbean are usually the ones that threaten this coast. Their paths are fairly predictable, for, as they curve slightly toward the northwest, they are confined by the shores of Central America and southern Cuba. It is unlikely they will start to recurve before they reach lat. 30°N.

The National Weather Service reports that tropical storms are most likely to occur in the gulf in late August and September with a secondary maximum in June. June storms usually form in the gulf rather than move in from the Atlantic or Caribbean.

WATERSPOUTS

Funnel clouds, which often cause waterspout activity, are commonly observed from these coastal areas and are most evident in July and August. The favored areas for common waterspout activity are large bays and inland waterways where shallow waters are heated by the hot summer sun, such as Tampa Bay and Pensacola Bay, Fla., Mobile Bay, Ala., the Louisiana delta, and along much of the northern section of the Texas coast.

Waterspout activity of tornadic origin can be expected to increase in February and subside after May on the Gulf of Mexico or adjacent waters. This is the same seasonal pattern that is projected by the National Weather Service for tornado frequency in the Central Gulf states.

NOOA'S WEATHER BUOYS

In 1972, the first of NOAA's monster weather buoys in the Gulf of Mexico was anchored about 195 nautical miles southeast of Gulfport, Miss.; later another was anchored approximately 175 nautical miles east of Brownsville, Texas.

These remarkable floating electronic sleuths are credited with improving the accuracy of gulf coast weather forecasting since their inception. The first weather buoy located in the gulf was described in a NOAA release as follows:

> This massive buoy is designed to withstand severe weather and sea conditions, including 150-knot hurricane winds, 60-foot (18-meter) waves, and 10-knot currents. Its platform carries a meteorological sensor package at levels of 15 and 30 feet (5 to 9 meters), a hull-mounted oceanographic sensor package, and 12 oceanographic sensor packages at various levels down to 1500 feet (460 meters). These packages will sense more than 76 individual measurements of environmental data during a routine weather reporting cycle.

The fact that merchant shipping in the Gulf of Mexico has suffered seriously from unpredicted severe frontal storms points up the need for additional buoys effectively located in the gulf to monitor the extension of cold-front lines into this region. These will be constructed as funding permits.

SECTIONAL CHARACTERISTICS OF THE GULF COAST

The nautical mileages for the shores of the states that border on the Gulf of Mexico, adapted from NOAA figures, are given as follows:

	General Coastline	*Tidal Shoreline*
Florida (gulf coast only)	669	4,427
Alabama	46	527
Mississippi	38	312
Louisiana	345	6,709
Texas	319	2,919
Total nautical mileage	1,417	14,894

Since the gulf coast of the United States falls within a span of about five degrees of latitude, the weather patterns are relatively uniform for the entire region. Our sectional divisions for this coast, therefore, are not based on differences in weather conditions. Instead, this section is divided geographically as follows: the Okeechobee Waterway; the west coast of Florida and its Intracoastal Waterway; the Gulf Intracoastal Waterway, East; the Mississippi River; and the Gulf Intracoastal Waterway, West.

The entire Gulf Intracoastal Waterway stretches from its eastern terminus at Carabelle, Fla., to Brownsville, Texas, for a distance of 920 nautical miles. The Waterway Charts are divided into east (E) and west (W) sections with mileage (M) 0.0 for both routes at Harvey Locks, New Orleans, La. The eastern section is 325 nautical miles long and includes the tidal waters of Florida's panhandle, Alabama, and Mississippi. The western section covers a distance of 595 nautical miles as it hugs the shores of Louisiana and Texas.

Prior to discussing the gulf coast itself, we feel it is important to include information on the Okeechobee Waterway in Florida, which is used by many persons to travel between the gulf and the Atlantic coasts.

Okeechobee Waterway The eastern entrance of the Okeechobee Waterway is situated approximately one-third of the way up the eastern side of the Florida peninsula from its tip. This protected waterway runs for about 130 nautical miles in a generally westerly direction from the junction of the St. Lucie Inlet and the Atlantic Intracoastal Waterway to the southern end of Florida's West Coast Waterway where the Caloosahatchee River empties into San Carlos Bay in the Gulf of Mexico. The Okeechobee begins its journey along the St. Lucie River and is routed through Lake Okeechobee, Florida's largest body of fresh water, from which it gets its name. This scenic inland waterway cuts through cattle country, citrus groves, sugarcane fields, and lonely miles of wooded lowlands.

Once you enter Lake Okeechobee, you are confronted with a choice of two routes. The decision concerning the best one to take should be based on the draft of your boat as well as predicted weather conditions. The shallow waters of Lake Okeechobee are likely to become rough when it is windy.

Route #1, which runs for about 21 nautical miles across the open lake waters, requires following a compass course. The soundings on the chart for this route are between a project depth of 2.4 meters (8 feet) to a maximum of about 4.3 meters (14 feet) although extended periods of either drought or excess rains can change the reported depths.

Even though Route #2 is 9 nautical miles longer, it is more popular because it is better protected. It has a project depth of 1.8 meters (6 feet) and is well marked. Barrier islands shield this route for about 13 nautical miles of its length. A 12-nautical-mile section close to the lake's eastern shore, however, is exposed to winds from the northwest and should be avoided whenever the wind blows from this direction.

Before beginning a journey on the Okeechobee Waterway, it is wise to telephone the Corps of Engineers at Clewiston, Fla., for the latest information regarding water depths and the operational status of the four locks that will be encountered, as these locks are sometimes closed for overhaul or repairs. There is a lift railroad bridge with a charted height of 15 meters (48 feet) less than 1 nautical mile east of the lake, so information concerning the water level is essential for those cruising aboard large auxiliaries.

Thunderstorm activity can be expected in this area, especially during the summer. According to a thunderstorm chart compiled by NOAA, a small section of southern Florida near Lake Okeechobee is the only place in the United States where over 100 thunderstorm days can be expected to occur each year. Radiation or ground fog also occurs frequently here on calm mornings and lasts for an hour or more; it is more prevalent along the section of the waterway west of Lake Okeechobee.

When you are heading east in the morning or west in the afternoon along this inland route, the tropical sun is glaringly in front of you. Dark sunglasses and wide-brimmed or long-billed hats are wardrobe essentials.

The 130-nautical-mile length of the Okeechobee Waterway is one leg of the approximately 430- to 475-nautical-mile-long Great Circle Cruise that circumnavigates southern Florida. The portion of the cruise on the eastern side of Florida covers from 170 to 215 nautical miles depending on which passage is selected between the Florida Keys to reach the Gulf of Mexico. The Great Circle is completed after a cruise of about 130 nautical miles in the Gulf of Mexico along Florida's west coast to the western end of the Okeechobee Waterway.

The West Coast of Florida and Its Intracoastal Waterway Perhaps one of the tersest statements ever made about year-round weather conditions by an auxiliary sailor is the following one sent to us about the weather in this area: "We sail all months of the year; winter is worst, spring and fall are best, summer—doldrums."

As the peninsula of Florida projects into the global belt of trade winds, the prevailing winds along its coasts are generally from the east. With few exceptions, the winds off the west coast of Florida, therefore, are generally lighter than those experienced off the east coast since most of them blow across land before reaching the gulf. The wind is especially light here during the summer.

Thunderstorms are prevalent in the summer and hurricanes are a possibility at this time as well. Winter storms occasionally make their presence felt, too. Cruising friends with local weather know-how rely on the "sun dog" phenomenon as a visual warning of approaching unsettled weather. This weather sign occurs, they say, when a "rainbow-like spot is visible each side of the sun."

There are three distinct boating areas along Florida's west coast and these closely follow the land divisions given in the literature of the Chamber of Commerce: the *Islands Coast,* starting at the southern tip of the peninsula and ending at the beginning of this shore's Intracoastal Waterway; the *Sun Coast,* extending nearly the length of the Intracoastal; and the *Big Bend,* taking in the area that curves into the panhandle.

There are shoal waters off the sparsely settled swamplands of the Islands Coast. Here, the low mangrove-covered islands of the Everglades and the Ten Thousand Islands are so low that many are awash much of the time, especially during the spring rains. Depths are charted from 2 to 4 meters (7 to 13 feet) for as far as 10 nautical miles from shore. A boat under power here leaves a foamy white trail of churned-up silt in its wake. Extremely shallow draft boats are required for exploration between the mangrove islands. Most of the boating, however, is done just off the mainland shore.

For many years, the Townsends' friend Colonel Allan Crockett headquartered in Naples, which is the largest settlement on the Islands Coast. The weather along this section of the Gulf of Mexico, according to this ardent cruising buff, is rarely bad enough to force anyone inland. A careful weather check is necessary, however, before starting out because all but a few of the passes between islands and the inside passages between islands and the mainland are tricky. Their charted depths are not reliable and they should be used only with local knowledge aboard. The openings from the gulf into the inside passages are often so small or obliquely oriented that they are invisible at 90 meters (300 feet).

Windy or rainy weather obscures the different colors of the water, one of the most important navigational aids in this area. The choppy seas caused by windy conditions build up quickly in these shallow waters.

The Sun Coast, next on the list of Chamber of Commerce west coast treats, is a delightful area for water sports, offering as it does the protected waters of the Florida West Coast Waterway and its easy access to the Gulf of Mexico; this part of the waterway is shielded from the gulf by barrier islands, and there are many well-marked channels between them. The mainland side of the waterway is highly developed with commercial and residential areas.

Despite the buoyed channels, this is an area of sandbars and shoals. Newcomers are urged to use the passes for the first time in clear weather with the sun overhead so that the darker color of the deeper water shows up. In this area, the gulf is deeper in comparison to the waters to the south, and, consequently, boating activity is noticeably greater.

The heart of the Sun Coast is large and important Tampa Bay. This sizable shipping and boating center, however, breeds its own weather peculiarities. The *U.S. Coast Pilot* especially warns of fog here. As this condition so rarely occurs south of Tampa, fog can take first-time boating visitors by surprise. During the Townsends' initial trip across Tampa Bay, they commented to each other on what they thought to be smoke from fires on the barrier islands. They began to realize it was fog when the trees started to disappear and they immediately verified their compass course with visual bearings before it enveloped them.

The Tampa Bay area has an excessive number of thunderstorms during the summer months. The local weather station records more storms in this vicinity than occur anywhere else along the entire U.S. Gulf of Mexico coast. Thunderheads build up well in advance of the storms, and these weather clues should not be ignored because the torrential rains usually associated with the storms quickly obscure all lankmarks and navigational aids.

Florida's West Coast Waterway has some piloting problems that are rarely present along the Eastern Intracoastal. Piloting is difficult in low visibility because this waterway is narrow and often lined with spoil areas—the long banks of sand resulting from the channel-dredging operations so essential to keep the waterway navigable. These spoil areas are either partially exposed or just below the water's surface and are in or near the edge of the dredged channel with the channel markers set as much as 6 meters (20 feet) outside of them. This marker setback presents serious difficulties when running compass courses in fog. The shoal areas make it risky to get close enough to a marker for positive identification of its number and to use it as a departure point for a compass course to the next marker. The great scarcity of anchorages off the waterway for boats drawing 1.8 meters (6 feet) or more sometimes makes it necessary to continue on in spite of a thick fog.

The string of barrier islands protecting the 130 nautical miles of this waterway ends at Anclote Keys. From here to the Gulf Intracoastal Waterway, East, there is an open-gulf trip across the Big Bend portion of Florida's west coast. This passage is called the "Missing Link" by waterway travelers accustomed to cruising the vast and nearly continuous chain of protected waterway routes along the major portions of the Atlantic Ocean and Gulf of Mexico coasts of the United States as well as the numerous connecting inland routes.

The most direct course across the gulf to another protected waterway is about 135 nautical miles, but most boaters prefer to add another 10 miles to the trip in order to use Cedar Keys as a visual check point. It is also becoming increasingly popular to port-hop between the few scattered harbors of refuge on the Big Bend, particularly since the fairly recent installation of six navigational buoys located to encourage this. Port-hopping can increase the distance of a cruise along this area to over 170 nautical miles.

The Missing Link passage is frequently used by gulf coast cruisers from the west as well as those from the Great Lakes who travel south in quest of constant warm weather during the winter. Since the wind usually blows from the south in this region, a southbound cruise often requires heading directly into the wind and seas, and consequently the going can become quite rough. Winter northers can produce such stormy conditions here that a trip can be dangerous even with a favorable wind. The best time for this southbound migration is late October or early November, after the hurricane season and before the beginning of the close-procession winter storms or northers. The trip north in the spring usually offers much smoother cruising with lighter winds generally from the south although a few days of northerly winds can be expected.

Haze or even heavy fog can develop here, sometimes unexpectedly, especially during the winter months or in the early spring. Cautious, accurate piloting is a must on this passage as proven by the experiences of two powerboat skippers on two separate occasions. In each case, they told us they called the National Weather Service forecast office in northern Florida before embarking on what they estimated to be about a 10-hour cruise for their boats across the Missing Link. The weather reports they received were excellent with no signs of storms or fog in the foreseeable future. Thick fog caught them both unawares and each admitted he had only a vague idea of the position of his boat. One skipper inched his boat in toward shore and anchored without incident. The other admitted that he followed the same course of action but struck a reef off Cedar Keys and damaged his boat quite badly.

Those experienced in making this passage suggest using an electronic depth indicator to steer along a fathom line, usually the 3-fathom line (5.5 meters or 18 feet) because of its proximity to navigational buoys, instead of following a compass course. They refer to this as "riding the depth finder."

Gulf Intracoastal Waterway, East The eastern section of the Gulf Intracoastal is about 325 nautical miles from Carabelle, Fla., to the 0.0 mile measurement at Harvey Locks, New Orleans. More than half of this section runs along Florida's panhandle with the rest bordering Alabama and Mississippi.

Gulf Intracoastal, East, is channeled through protected waters considerably wider and deeper than the majority of U.S. intracoastal routes. It is protected from the gulf by barrier islands as far as 5 to 10 nautical miles from the mainland. As a result, the majority of pleasure boaters bent on fishing or a day's spin on the water are likely to pursue their sports within the confines of the barrier islands or on one of the broad bays or sounds through which the waterway is threaded, rather than on the more open waters of the Gulf of Mexico.

Commercial traffic is heavy on this waterway, as it serves as an artery for transporting petroleum, chemicals, and other types of cargo between the larger southern ports and the Midwest or Great Lakes via the Mississippi River.

A study of the general weather picture for the area as conducted by NOAA reveals that 15 to 20 frontal systems can be expected each year. Rainfall is usually in the category of showers and is either associated with thunderstorm activity or fronts. The prevailing winds are southerly from March through July, easterly in August and September, and northerly from October through February. The colder months, November through April, have the highest incidence of fog for the year, with fog occurring with greatest frequency in March.

Virginia (Winky) Grandison, one of our recreational boating sources of local knowledge and a close friend of Sallie's, advises newcomers to this beautiful section of the coast to be wary, weatherwise, particularly in the hot and airless summers, as weather changes can be swift and violent. Heavy rain and severe thunder and lightning storms arrive almost unannounced to punctuate the usual light air wind pattern. September and October are good months for cruising and fishing, with generally pleasant and cooler weather, although once again storms strike with very little visual warning. November through February can bring cold weather despite the southerly location; Winky recalls being surprised by a low temperature of 17°F (−8°C) shortly after she moved to the area. Rain can be expected often during the winter season, especially during January and February. March through June are the best months for boating, and although some fog can be expected, it usually occurs at night. The greatest weather hazard during the spring is the threat of tornado activity. The local radio stations, however, can be depended on for fine tornado reporting.

Mississippi River By the time this great river reaches the Gulf of Mexico, it has provided drainage for more than 3 224 550 square kilometers (1,245,000 square miles), an area that includes all or part of 31 states of the United States and two provinces of Canada. The Mississippi River begins its journey to the gulf as an unpretentious brook in tiny Lake Itasca in Minnesota. From there, it eventually develops into a vast network of inland navigable waterways to form an astounding river system of about 19 875 kilometers (12,350 miles) in length. The famed "Ole Man River" in conjunction with the Illinois Waterway and the Chicago Ship Canal also serves as the connecting link between the Great Lakes and the Gulf of Mexico. This waterway touches seven states on its journey from Lake Michigan to the gulf—Illinois, Missouri, Kentucky, Tennessee, Arkansas, Mississippi, and Louisiana.

While the river is primarily considered as an artery for commercial traffic, those with recreational boats on the Great Lakes use it to reach the Gulf Intracoastal. There is a section in the Corps of Engineers publication *Mississippi River Navigation,** entitled "For Part-time Pilots," specifically intended for recreational boating. Following is a brief synopsis of river hazards to be on the lookout for during a downstream trip to the gulf gleaned from this and other government publications.

> On the upper Mississippi, there are some very large bodies of water through which you will pass as you head down the river. Wave action created by strong winds on these confined waters can reach such proportions as to make travel by small craft hazardous.
>
> Fog occurs infrequently north of New Orleans and usually only during the winter months. However, it may be encountered anywhere along the river from New Orleans south to the gulf, a distance of about 160 kilometers (100 miles). Southerly and easterly winds bring it in and northerly and westerly winds clear it away.
>
> There are two rainy periods; one of localized, scattered summer showers occurring from about mid-June to mid-September, and another of slow, continuous winter rains, often lasting for several days, occurring from mid-September to mid-March.

* Order from: Corps of Engineers, Vicksburg, Miss. 39180

The speed of the current in the river averages from approximately three to five knots. During the peak of the high water stage, the speed of the current is even greater. During the low or medium water stage, the river current speed ranges from nearly two to three knots. Bankfull stages are likely to occur on the lower Mississippi anytime from December through July and most frequently in March or April. Water levels in the river are lowest in the fall, particularly in October and November. This is the time to be on the lookout for sandbars.

Floating ice is a hazard in the upper portions of the river from early January through about mid-February as a result of freezing weather conditions to the north during these months. During extremely cold winters, ice floes have been spotted in the river as far south as Natchez, Miss., with small chunks of floating ice seen bobbing by White Castle, La.

The hazard to boating most often associated with the Mississippi by readers of Mark Twain's river yarns is the preponderance of floating or partially submerged debris. Mr. Twain's experience all those years ago is still valid today because this debris, which includes tree trunks and limbs, is still likely to be encountered in the river throughout the year. Operators of small craft are advised to maintain a sharp lookout at all times and night travel by pleasure craft is not recommended.

Sufficient motor power is required aboard all boats on the river so that they can get out of the current to provide room for large commercial vessels and tows. Barges with tows that frequently exceed 305 meters (1,000 feet) in length and may be as much as 76 meters (250 feet) wide ply the Mississippi River. The average modern steam or diesel towboats may push as many as twenty 907-metric-ton (1,000-ton) steel barges at one time. When underway large vessels and tows cannot be maneuvered easily or stopped quickly and they should always be given as wide berth as possible—*it takes from a half mile to a mile to bring a tow to a stop*. Ocean-going ships also navigate in the lower portion of the river.

Above Baton Rouge, La., a depth of 2.7 meters (9 feet) is available in the Mississippi River and the Illinois Waterway to Chicago and Lake Michigan. Limiting clearances in the Illinois Waterway are: bridges, 23 meters (75 feet) wide and 5 meters (17 feet) high; locks, 33 meters (110 feet) wide and 183 meters (600 feet) long.

The lockmaster is charged with the immediate control and management of locks. The lockage of pleasure craft is expedited by locking them through in company with commercial craft, with the exception of barges carrying petroleum or other highly hazardous material, in order to utilize the capacity of locks to their

> maximum. If, after the arrival of pleasure craft, a separate or combined lockage cannot be accomplished within a reasonable time (not to exceed the time required for three other lockages), separate lockage will be made.

Because of the many problems inherent in long-distance cruising on the river, many Great Lakes boat owners prefer to have their boats shipped by barge. Many companies in the vicinity of the Illinois Waterway offer this service.

Gulf Intracoastal, West From mile 0.0 at Harvey Locks in New Orleans, the Gulf Intracoastal Waterway meanders to the west inland through the bayou country of Louisiana and then swings back to follow the coast in Texas. The Texas section runs inside very long, low barrier islands to its terminus at Brownsville, Texas. This entire western section of the Gulf Intracoastal covers a distance of 595 nautical miles.

Oil derricks and other structures in the gulf are testimony to the development and utilization of oil and gas wells off this portion of the coast. Visible signs of the sprawling structures required by this industry are seen ashore; in bays, lakes, and lowlands along the shore; as well as in the offshore waters.

Designated shipping lanes are called fairways, and drilling apparatus is not allowed in these lanes. They are marked on navigational charts and are safe routes to take through the oil fields, but there are no navigational aids provided for the courses. The lights and fog signals on the commercial structures outside fairways are given in the *U.S. Coast Pilot*. Selected weather information for the continuous weather broadcasts from U.S. Coast Guard stations is supplemented by observers on oil rigs in the Gulf of Mexico.

Since the Intracoastal Waterway, West is commercially oriented, we have drawn heavily on the local knowledge of many of our recreational boating colleagues for advice for newcomers to this area. A cruising couple, Lou and Henry Finke of Louisiana, wrote us a most informative report under their letterhead, "Aboard *Stardust*":

> This is a bit of sound advice for cruising the Gulf Waterway: it is primarily for the movement of barge traffic and is heavily used. In general, the waterway is narrow; the tows are long and heavily laden; the tugs have all that they can do to keep their barges in line. To us, they have been polite and courteous, but we always give them the largest part of the channel and the right of way; after all, we are cruising for fun, they are working.

Pat Patterson of Texas, a former racing competitor of the Townsends, echoes the Finkes' warning, adding the admonition to anchor well off the waterway at night because many have been killed who have not followed this precaution. A. E. Chester, also of Texas, notes that the locks on the Gulf Intracoastal have heavy barge traffic, which as you know has priority, and he recommends making radio contact with the lockmaster to determine your locking order.

Weatherwise, the best time for boating is from late March through November, although occasional northers can be expected during this time of the year. According to the Finkes, "As a norther approaches, the wind goes through a clockwise cycle, going from east, through southeast, then southwest, finally northwest and then north when the weather will clear up and be beautiful for several days with the wind out of the northeast."

Slow, continuous winter rains occur from mid-September to mid-March and are generally associated with frontal activity. The greatest incidence of fog is experienced during the winter and very early spring, and tornadoes and waterspouts of tornadic origin may occur as well. The prevailing wind is from the southeast for most of the year and is stronger during the winter.

During the summer, the humidity is especially oppressive in the Louisiana portion of the waterway, and this increased humidity breeds thunderstorms. "In the middle of almost every very hot summer afternoon," the Finkes' letter continues,

> cumulonimbus clouds build up, bringing thundershowers and even thunderstorms, some of which are quite severe. These can bring winds for 15 minutes to a half hour of up to 70 knots, with an 180° wind shift as they pass. Cumulonimbus afternoon thunderstorms are frequent, but local knowledge is available so that one can learn how severe to expect them to be. The rule is that the darker the cloud and the higher it towers, the more severe the storm. We have never experienced strong winds without dark clouds. During the months of July and August, there is a thunderstorm almost every day, most of which are mild. As we said previously, one can soon learn the severe ones in time to take down sails and to anchor. They last 20 to 30 minutes so no sea is built up. Hurricanes are well advertised and when one is on the way there are many safe hiding places up rivers in the pine forests.

Compared with the summer thunderstorm activity on the coast of Louisiana, there is a decided decrease in this type of activity along the coast of Texas. However, the average velocity of the wind is greater on this most westerly portion of the waterway, and often the summer wind is hot and dry because of its proximity to the plains of Texas. Here, the old weather saw is transposed to, "It's not the humidity, it's the heat." Whenever the wind blows from the southwest or west during this season, it brings in the scorching heat of the plains to the coastal area of Texas.

Summertime is the height of the boating season in the gulf in spite of the heat. Many of the marinas and yacht clubs have covered docks, not only to protect the painted and varnished surfaces of the boats, but also to provide the boaters some relief from the relentless sun. More and more cruising boats in this area have air conditioning. A hat is an essential item of boating attire.

One soon learns that a deck painted in any other color than white is uncomfortable to sit on. Even a light gray deck is hot and to sit on a dark color is to risk a scorched derrière.

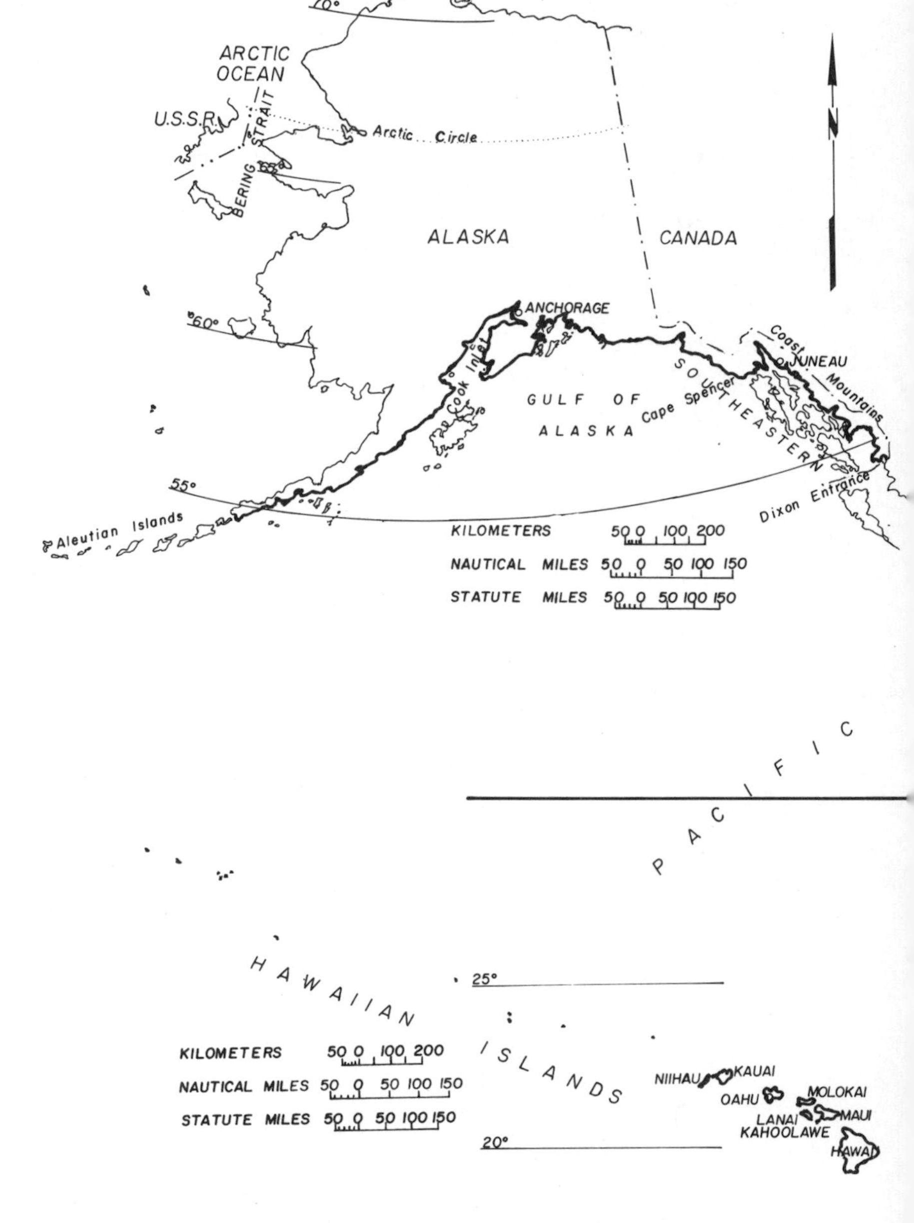

Fig. 20. THE PACIFIC COAST—Including Alaska and Hawaii

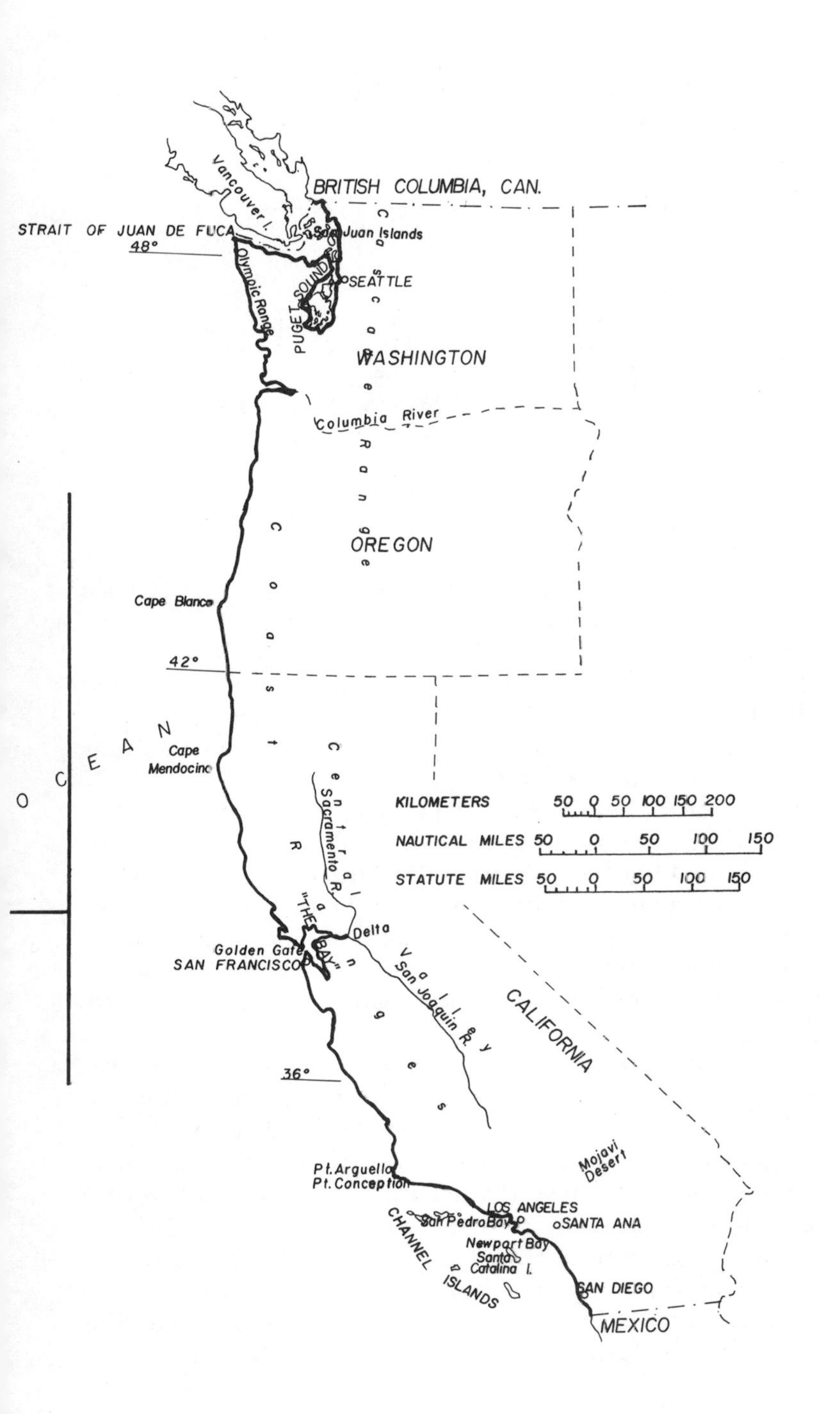
BRITISH COLUMBIA, CAN.
Vancouver I.
STRAIT OF JUAN DE FUCA
48°
San Juan Islands
SEATTLE
Olympic Range
PUGET SOUND
WASHINGTON
Cascade Range
Columbia River
OREGON
Cape Blanco
42°
Coast Ranges
OCEAN
Cape Mendocino
Central Valley
Sacramento R.
KILOMETERS 50 0 50 100 150 200
NAUTICAL MILES 50 0 50 100 150
STATUTE MILES 50 0 50 100 150
"THE BAY"
Delta
Golden Gate
SAN FRANCISCO
San Joaquin R.
CALIFORNIA
36°
Mojavi Desert
Pt. Arguello
Pt. Conception
LOS ANGELES
San Pedro Bay
SANTA ANA
CHANNEL ISLANDS
Newport Bay
Santa Catalina I.
SAN DIEGO
MEXICO

CHAPTER 11

Pacific Coast Weather (including Alaska and Hawaii)

Horace Greeley's famous and much repeated advice in the 1840s to "go West, young man, go West!" is still heeded by many today as the fabled, glamorous, exciting Pacific shores of the United States continue to exert magnetic appeal. Recreational boaters have responded to this lure by the thousands in recent years, and it has been a struggle in the more densely populated areas to expand boating facilities quickly enough to accommodate them.

The Pacific Ocean, which still has many of its secrets locked in its depths, is the largest ocean in the world. Holding half of the water on earth, it touches more of the shores of the United States than twice the combined measurements of the country's Atlantic Ocean and Gulf of Mexico general coastlines. Along most of the U.S. Pacific coastlines, the waves of this great ocean break against rocky escarpments rising abruptly from the waters' edge against a backdrop of lofty mountain peaks.

Weather forecasters in the states along the continental seacoast, as well as in Hawaii some 1,700 nautical miles out in the ocean, are handicapped by the scarcity of land-based weather stations to the west to warn of approaching weather systems. Meteorologists in these areas rely on satellite pictures and observations from ships and airplanes for most of the material compiled for their forecasts. Of necessity, the Cooperative Ship Program is very active in the Pacific. A representative of the National Weather Service told us that an amazingly large number of weather reports are received from aircraft in flight over the Pacific Ocean as well. Offshore weather buoys also play an increasingly important role in recording and transmitting meteorological data. Observing stations located both in the United States and in other countries are another source for weather information.

THE BATTLE OF THE PRESSURE SYSTEMS

The weather patterns affecting our Pacific coasts are dominated by two massive circulations of air that vie with each other during the course of a year. The Pacific High holds sway from spring to fall and the Aleutian Low exerts the major influence during the winter months.

The center of the Pacific High is located northeast of the Hawaiian Islands in the summer, approximately 700 to 900 nautical miles due west of San Francisco. At this time, it blocks most of the normal west-to-east progression of weather masses moving toward the west coast of the continental United States.

The influence of the Aleutian Low, which resides over the 870-nautical-mile-long Aleutian Islands chain in the wintertime, subsides in June to become almost nonexistent through the summer. A procession of lows begins to move through this semipermanent trough of low pressure in September, reaching maximum activity during the winter months.

SURFACE WIND

The Pacific High sends out a huge clockwise circulation of air that is diverted from its natural path by the mountain ranges along the mainland coast. This air flow or surface wind is forced, instead, to flow down the coast where it more or less follows the natural contours of the land. It often increases in velocity to gust around the high bluffs. A gap in the Coast Mountains, such as one caused by a river flowing into the Pacific, releases this wind from its confining wall of rock to blow with increased intensity through what is probably best described as a natural funnel.

During the summer boating season, the velocity and direction of surface wind is influenced by the location of the Pacific High in relation to the location and configuration of land masses. Thus, the northern shores of the mainland experience prevailing westerly winds and the southern shores, prevailing northwesterly winds.

In the Hawaiian Islands, the summer air circulation of the Pacific High joins with the east-to-northeast circulation of the trade winds and therefore increases the velocity of the prevailing winds at this time.

FOG

In contrast to the year-round fog-free weather experienced in Hawaii, no month is free of fog along the Pacific coast of the mainland. Since the Pacific High is centralized over waters that are comparatively warmer than those along this coast, the warm, moist air flowing from it over the cooler coastal waters frequently causes advection or sea fog to form. This fog is especially prevalent in the summer and often lies in a thick bank near the coast during the day. It rolls inshore about sundown and generally retreats from the coast in the morning.

Fog does not occur off this coast as frequently in the winter, but when it does it is often more dense than summer fog. Winter fog in this region is primarily radiation fog. It forms in the lowlands of the coastal valleys and protected harbors when moist air is cooled to the dew point at night. The *U.S. Coast Pilot* comments on winter fog as follows:

> It is more local in character and although it may extend over a considerable range in latitude it seldom extends any great distance to sea. However, when the so-called summer or advection type of fog, which may also occur in winter, unites with fog which has formed over the land, a sheet of fog may extend a considerable distance to sea.

Industry and forest fires also contribute to the lowering of visibility along the coast. Smoke from stacks in heavily populated areas pollutes the air on occasion, causing a dirty yellow smog to blanket cities and nearby coastal areas. The heavily forested areas of the coast are subject to forest fires in the summer when rains are infrequent. They are often caused by lightning strikes in the mountains, although they are also started by campers carelessly leaving inadequately smothered campfires. The gray-brown smoke from forest fires has been known to lower the visibility on nearby coastal waters hazardously, but this condition is not likely to occur more than a few days each year.

Depending on the type of electronic or other equipment aboard, the high land along these coastal waters can be considered a mixed blessing when navigating in a fog. When using a radio direction finder, there are times when a radiobeacon signal is deflected by steep bluffs and sometimes even blanked out entirely, rendering the instrument ineffectual for piloting. On the other hand, the reflection on a radarscope of predominant and often distinctive outlines of this shore aid in shoreline identification. The rugged, high coastline also offers the possibility of using echo sounding as a navigational tool as described in the *U.S. Coast Pilot:*

> In foggy weather, the distance offshore frequently can be estimated by noting the elapsed time between the sounding of a ship's whistle or siren and the resultant echo from the sides of hills or mountains. The distance in nautical miles from the hill or mountain is about one-tenth the number of seconds between sound and echo. In narrow channels with steep shores a vessel can be kept in midchannel by navigating so that echoes from both shores return at the same instant.

OCEAN CURRENTS

The chilly waters present adjacent to the western coast of the continental United States are caused by the presence of cold northern currents. The North Pacific Current flows in an easterly direction to an area off Seattle, Wash., where it splits into two branches called the California Current and the Alaskan Current.

The California Current flows south under the warmer coastal waters, which are pushed away from the shoreline by the prevailing winds that parallel this coast. The cold undercurrent thus is enabled to hold sway near the coast. This condition is referred to as upwelling.

The Alaskan Current curves up into the Gulf of Alaska. Sometimes it is joined and chilled by frigid waters from the Arctic Ocean. The area where these waters meet is called the Arctic Convergence.

STORMS

Severe winter storms resulting from the weather triggered by the migration of lows through the Aleutian Low causes most boaters in northern California, Oregon, Washington, and Alaska to confine their boating season to late spring, summer, and early fall. In southern California and Hawaii, where the boating season is year-round, the winter storms of the northern climes are usually apparent only because of an increase in ocean swells.

Our Pacific coasts are singularly free of thunderstorm activity. A Thunderstorm Chart compiled by NOAA depicts the coastal areas of Washington, Oregon, and California to be in a "0 to 10" zone of thunderstorm days per year. A notation on the chart adds that Alaska and Hawaii average less than 10 thunderstorm days per year.

These regions, therefore, are comparatively free of waterspouts of tornadic origin, although common waterspout sightings have been reported off the coast of southern California almost every year. The tornadic waterspouts that occasionally occur off the Pacific coasts of the United States are more prevalent off the northern coast of the mainland.

Upon rare occasion, a tropical storm strikes a Pacific shore of the United States. The occurrence of one in other places in the Pacific Ocean, however, is advertised by the long swells that outrun the storm.

SWELLS

Large, long swells driven by the prevailing winds undulate across the broad expanse of the Pacific Ocean. They increase in size and frequency as they near a coast whenever a sea breeze adds to the velocity of the prevailing winds. (Conversely, when the wind blows from the land toward the open ocean, swells tend to diminish.)

High swells cause discomfort and even danger to persons aboard boats riding at anchor. Since this is a great problem in this region, fishermen and others have been compelled to contrive devices to dampen the motion of their anchored boats. Bilge keels or stabilizing fins, permanently affixed to the bottoms of boats, are sometimes used to moderate the disturbing action of the swells.

One portable device has the delightful name of flopper stopper and is frequently homemade. As Easterners, we find it so intriguing that we couldn't resist including a description of one shown to us by Pat Guerlach, a West Coast boating friend. A flopper stopper constructed according to the following measurements will be of sufficient size for a 6.7-meters-long (22-foot) boat, but can be adapted to any size boat.

This flopper stopper consists of a 3.8 centimeters (1½ inch) aluminum frame that is about 0.8 meter (2 feet, 6 inches) long and 0.6 meter (2 feet) wide. (Frames made of plywood will work successfully if they are weighted.) Hinged to the longer sides are two doors that open upward.

The device is suspended by all four corners from a bridle attached to a line on an outrigger that projects over one side of the boat. The flopper stopper is lowered to about 1 meter (3 feet) below the surface of the water. As the boat rolls toward the flopper stopper, the frame sinks and the doors open. When the boat begins its righting motion, the frame rises and the doors close, thus dampening the boat's rolling motion considerably.

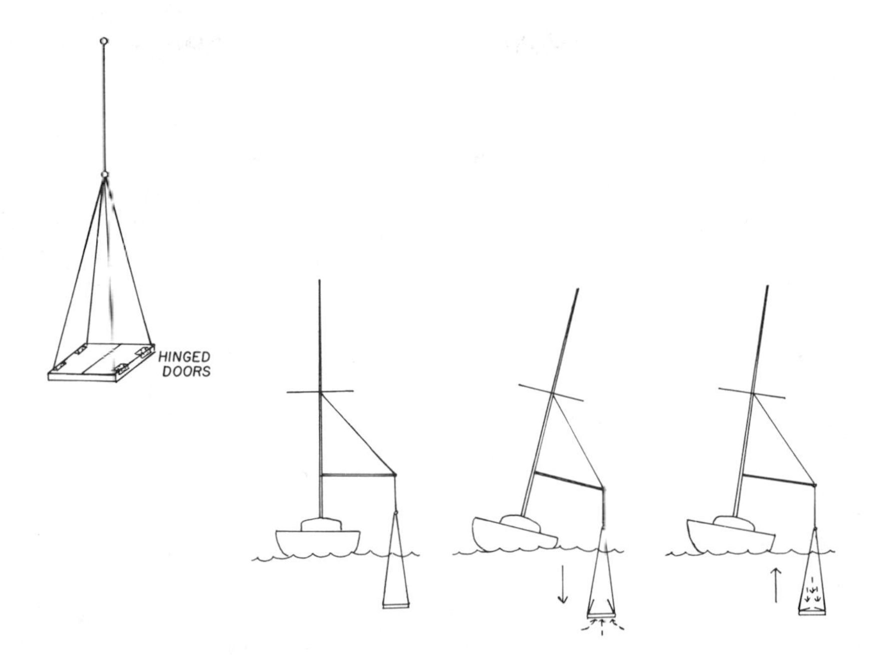

Fig. 21. FLOPPER STOPPER *Hinged doors on a frame are pressed open by the water as the boat rolls toward the flopper stopper and close to retard the motion as the boat comes back to an upright position.*

KELP

A coarse, rubbery type of seaweed, kelp grows on nearly every rocky area beneath the surface of the coastal waters from Alaska to southern California. This sinuous growth anchors itself to an underwater rocky base and sprouts brownish-yellow blades or bladders that grow so long that they spread out to float on the water after they reach the surface.

Skippers avoid crossing kelp beds because the weed tends either to wrap itself around a turning propeller with the tenacity of strong rope or to clog the water intake of a salt-water-cooled engine. Strong currents may sweep the floating kelp fronds below the surface of the water to menace the unwary. During severe storms, kelp is wrenched from the rocky bases and snarled into free-floating seaweed islands. Kelp beds are more prevalent and widespread in the summer, as they are less likely to be uprooted by storms and high seas.

One of the largest concentrations of kelp in the Pacific Ocean is found off San Diego Bay, mute testimony to the rocky ocean floor in this area.

TSUNAMIS

A tsunami, pronounced "sue nah'mee," is a series of waves that can travel thousands of miles through the sea and erupt into one or more giant waves that overwhelm a shore with a destructive onrush of tons of turbulent water. This phenomenon usually consists of a sequence or train of 5 to 10 waves triggered by seismic submarine disturbances that cause dramatic changes in the contour of the ocean floor. These disturbances may be underwater earthquakes, submarine volcanic eruptions, or massive sea bottom landslides. Since tsunamis are caused by forces that are entirely different from those that cause astronomical tides, it is incorrect to call them tidal waves.

The danger of the occurrence of a tsunami is especially prevalent in the Pacific Ocean because about 80 percent of all seismic activity on earth takes place in this region. The Tsunami Warning System (TWS) in the Pacific is operated by the National Weather Service, with headquarters at Ewa Beach, near Honolulu, Hawaii. The system carries on the detective work necessary for the prediction of possible tsunami activity that may affect any of the shores of the Pacific Ocean. It also provides watch and warning information to countries and territories throughout the Pacific. In one of their publications, NOAA states: "Sooner or later, tsunamis visit every coastline in the Pacific. This means that the tsunami warnings apply to you if you live in *any* Pacific coastal area."

To date, no method has been discovered to predict accurately the size a tsunami may attain when it reaches a shore. However, it is possible to determine the approximate magnitude of an earthquake by means of an instrument called a seismograph.

According to excerpts from NOAA's *Tsunami Safety Rules:* "A small tsunami at one beach can be a giant a few miles away . . . . All tsunamis—like hurricanes—are potentially dangerous . . . . Never go down to the beach to watch for a tsunami; when you can see the wave you are too close to escape it."

In order to give even a simplified explanation of tsunami waves, it is necessary to discuss certain aspects of waves in greater detail than we did under Sea Conditions (page 35).

Waves generated by wind are short waves; tsunami waves generated by underwater disturbances are long waves. At sea, short waves are those previously described as deep-water waves, and their wavelength, or the distance between the crests of consecutive waves in a series, is shorter than the depth of the water under them. In the case of long waves, the wavelength is much greater than the depth of the ocean through which they travel.

Long and short waves differ in the vertical plane. From this aspect, although the water particles near the surface of a short wind wave move in an elliptical orbit as each wave passes through it, in a long wave or tsunami wave, the water is disturbed about equally from the surface to the ocean floor. The water particles move horizontally back and forth as a unit for the entire depth of the water. As the wave energy of the long wave excites the water particles all the way to the bottom of the sea, tsunamis are said to "feel the bottom." The surface speed of short waves may be calculated from their wavelength (1.35 × square root of the wavelength = wave speed in knots). The speed of long waves, however, is determined by the depth of the ocean through which they pass. Their speed increases as the water deepens. A tsunami or train of waves passes swiftly through surface short waves, leaving them behind. Long waves cover great distances of open ocean.

Because the speed of long waves is determined by the depth of the ocean, concentric lines of equal speed of travel can be drawn on a chart and thus the time of arrival of a tsunami at a particular coastal area can be estimated no matter how far away. This method of prediction has proved to be accurate within about 1½ minutes in an hour. Figure 22, a chart made available to us by NOAA, illustrates the projected arrival time at various places around the Pacific Ocean of an actual tsunami generated by an earthquake that occurred approximately in the middle of the chain of Alaska's Aleutian Islands.

The height of a long wave gives no indication of its speed. In the deep ocean, for example, a tsunami wave may measure only a foot or two* from trough to crest on the surface of the ocean even though it is traveling at a speed of more than 400 knots. The wavelength between tsunami crests is often more than 100 nautical miles. Thus, these waves will pass unnoticed by ships at sea and by aircraft flying overhead. Without the services of the Tsunami Warning System, a tsunami could arrive at a coast unannounced without causing any perceptible change in the short waves in the area until just before its series of waves rose up to crash on shore.

The potential threat of tsunamis starts as they near a coast be-

---

* 0.5 meter.

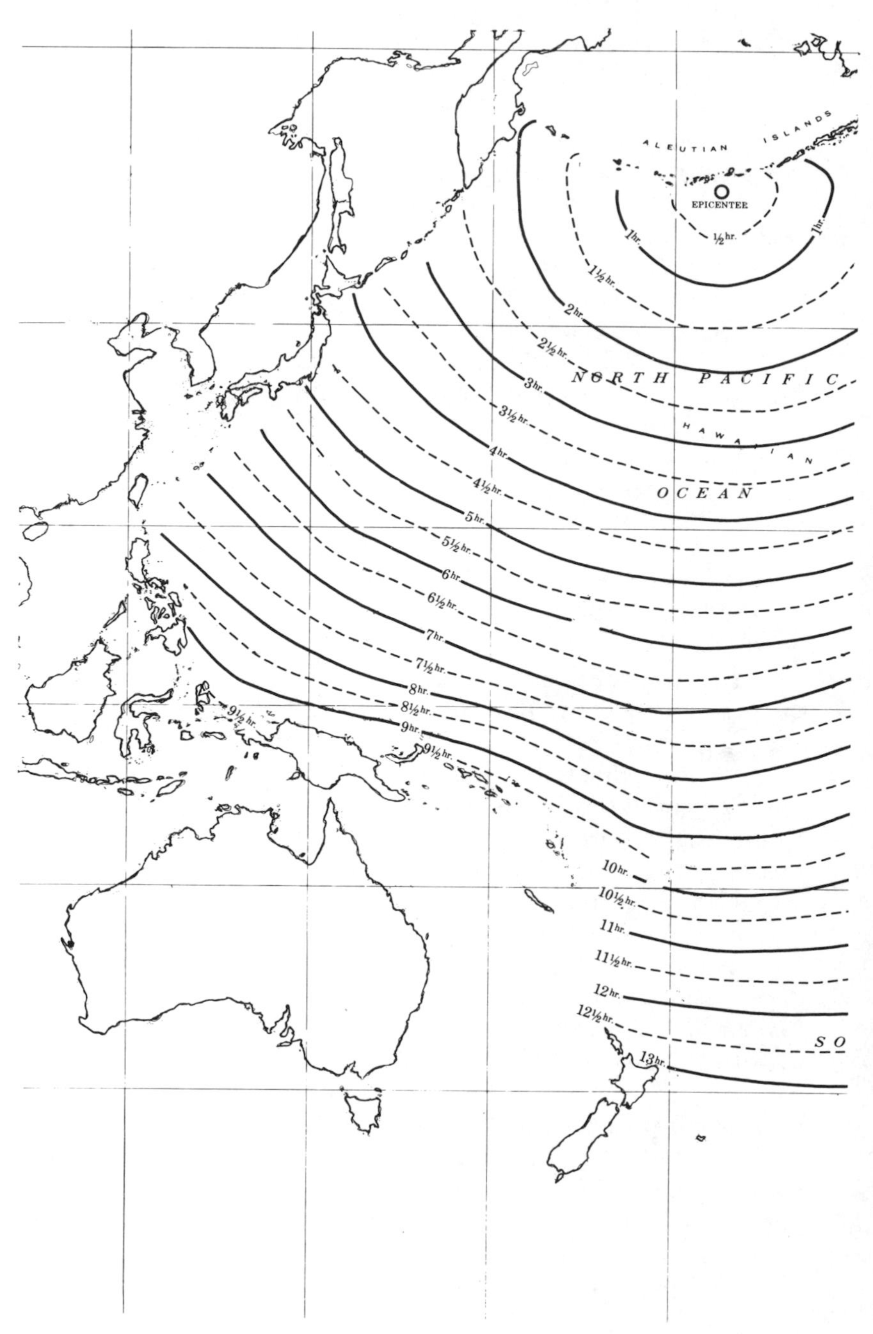

Fig. 22. TSUNAMI TRAVEL TIME CHART *Published by NOAA.*

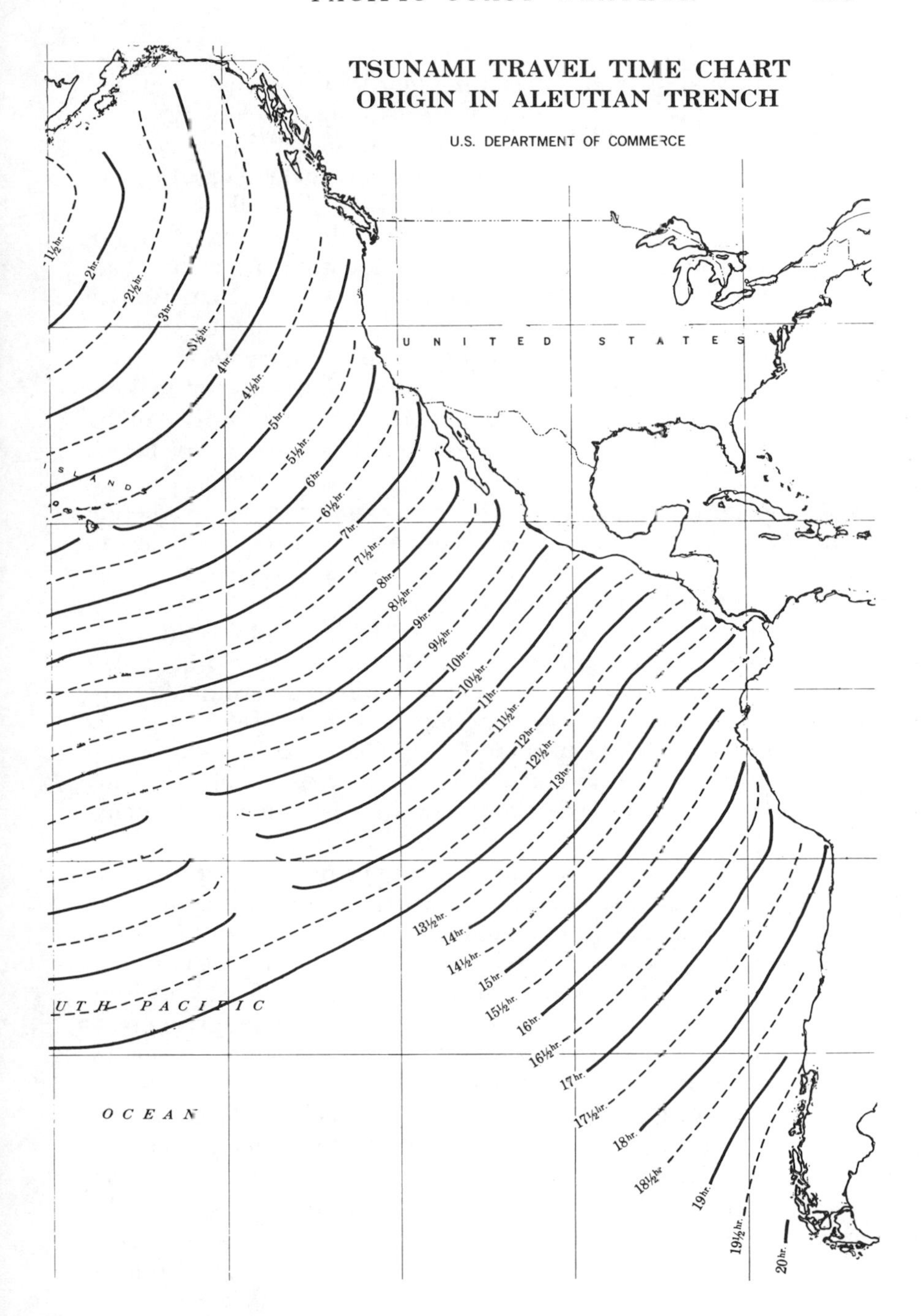
TSUNAMI TRAVEL TIME CHART
ORIGIN IN ALEUTIAN TRENCH
U.S. DEPARTMENT OF COMMERCE
UNITED STATES
1½hr.
2hr.
2½hr.
3hr.
3½hr.
4hr.
4½hr.
5hr.
5½hr.
6hr.
6½hr.
7hr.
7½hr.
8hr.
8½hr.
9hr.
9½hr.
10hr.
10½hr.
11hr.
11½hr.
12hr.
12½hr.
13hr.
13½hr.
14hr.
14½hr.
15hr.
15½hr.
16hr.
16½hr.
17hr.
17½hr.
18hr.
18½hr.
19hr.
19½hr.
20hr.
UTH PACIFIC
OCEAN

cause the ascending slope of the ocean floor around coastal areas alters the vertical wave form of all waves. Like short waves in shallow water, the long waves of tsunamis tend to increase in height as they near a shore, although the huge waves for which they are noted do not occur every time. A giant wave is said to be more likely to develop in size on gradual sloping shores than on steep rises of the ocean floor.

Sometimes, but not always, the first visible warning of an approaching tsunami is a marked withdrawal of water from the shore exposing rocks and shoals and happening so quickly that fish are stranded to flap about high and dry. Once the first tsunami wave has broken on shore, the danger is not necessarily over. Many persons have been killed by a later gigantic wave rising from a fairly normal-looking sea to explode against the shore. The period between the arrival of the crests may be 10 to 15 minutes or more. Usually the third or fourth wave in a tsunami train produces the infamous and destructive great wave. A monster wave such as this has been known to rise to a height of 30 meters (100 feet) as it swept over coastal areas or built up in V-shaped harbors or estuaries. The word tsunami comes from the Japanese and literally means harbor wave.

Fortunately, not all submarine earth tremors produce tsunamis, and those that do develop rarely reach destructive proportions. We understand that only about 10 percent of the larger earthquakes are tsunami producing. Reports over a period of years indicate that about five tsunamis occur each year in the Pacific Ocean, and most of these are of little consequence. For example, on the average, one destructive tsunami is likely to strike the Hawaiian Islands every twenty years.

Seismographs in Hawaii and many other places in the Pacific region record all earthquake activity. They provide a continuous visible seismic record, and most of them have an alarm that goes off when earth tremors are recorded. Reports from all stations are transmitted to Tsunami Warning System headquarters at the Honolulu Observatory where a 24-hour seismic watch is maintained. A seismograph operator is able to determine the surface distance of a quake from his station by determining the differences in travel times of the various types of seismic waves recorded on his seismograph. As each station reports its data to headquarters, circles of distance are drawn on a chart around each reporting station. These circles of distance all intersect at a surface location directly above the earthquake. This location is called the epicenter. The epicenter is then plotted on previously prepared charts that show tsunami travel time to various locations from any point in the Pacific. With the time of occurrence of the earthquake known, it then becomes simple to compute the arrival of the tsunami at various coastal areas around the Pacific Ocean.

Once a quake has been located, areas nearest to the epicenter are notified first of the location and magnitude of the disturbance; it usually takes less than one hour for this initial warning notification to be prepared. This initial notification is a tsunami watch.

Once the danger of a tsunami has been ascertained, a second warning phase is set into operation. Urgent calls from headquarters are sent to about 50 tide stations at island and coastal locations in the Pacific Ocean region, requesting operators to read and monitor their marigrams (tide gauge registers) and report back. A marigram will register a distinctive irregularity whenever it detects the presence of waves that have resulted from a seismic disturbance. When tide station operators report any unusual marigram activity, the Tsunami Warning System goes into high gear. Time is of the essence!

Foreign and domestic locations in the Pacific are immediately notified of the development of a tsunami, its predicted time of arrival in their area, and the applicable registered marigram heights. In the United States, the National Weather Service and the U.S. Coast Guard are among the federal agencies that relay this message to the general public. Radio, television, and newspapers then assist in spreading the warning. On local levels in areas that may be affected, police, fire, and civil defense personnel mobilize to evacuate residents from coastal areas. Captains of ships are alerted and advised to move out to the safety of the open ocean and, if possible, to at least a mile or two beyond a depth of 18 meters (60 feet). A state of emergency exists until the TWS is sure that there is no longer any danger.

The different phases of a tsunami alert are explained in a NOAA publication as follows:

> A tsunami watch is issued when the proper-category earthquake is detected. A tsunami warning is issued when the presence of seismic sea waves is confirmed. This one-two approach is the same procedure as NOAA uses for hurricanes and tornadoes—first an alert to *possible* danger, then a warning of *actual* danger. Included in the warning are quite precise forecasts of the arrival times for the first wave of the tsunami at populated points in and around the Pacific basin.

Even a warning of an approaching tsunami that is expected to be inconsequential must be heeded by those in a harbor because certain frequencies in the series of the wave train may cause a seiche similar to those that occur on the Great Lakes.

SECTIONAL CHARACTERISTICS OF THE PACIFIC COASTS

The Pacific Ocean breaks on the shores of California, Oregon, Washington, Alaska, and Hawaii in the United States. This is one of the few things Hawaii and Alaska have in common; from a weather point of view the difference between their climates is enormous, as the latitude of the Hawaiian Islands is a scant 19° north of the equator and the latitude of Alaska's peninsula on the northern edge of the Pacific Ocean is only 22° south of the North Pole. The remaining three Pacific coastal states enjoy their own weather differences, but these are considerably less extreme.

The nautical mileage of the general and tidal coastline of each state, adapted from NOAA figures, provides interesting comparisons between the five states, as follows:

	General Coastline	*Tidal Shoreline*
California	730	2,978
Oregon	257	1,225
Washington	136	2,630
Alaska (Pacific coast only)	4,849	27,271
Hawaii	652	914
Total nautical mileage	6,624	35,018

We divide the southern and northern portions of California in our weather discussion because of the geographic configuration that is responsible for a noticeable weather change along this state's coastline. The states of Oregon and Washington are handled together as they fall naturally into a single weather category. The information concerning Alaska is confined to its Pacific coast, and that concerning Hawaii to its eight high islands.

Southern California The Coast Mountain Range, situated anywhere from 32 to 64 kilometers (20 to 40 statute miles) inland from the ocean, parallels the rocky shores of California. The islands situated some distance from this coast are a continuation of the coastal mountain chain, although with these few exceptions the mountainous terrain is far below the surface of the ocean. Unlike the large portions of the Gulf of Mexico and the Atlantic Ocean coast that are protected by long stretches of barrier islands, this coastline generally is open to the long, long fetch of the Pacific Ocean. Thus, a great deal of the pleasure boating is concentrated on large bays in or near the cities of San Diego and Los Angeles. The few harbors along the coast were hard won—nature provided the barest beginnings and men did the rest by constructing breakwaters, diverting rivers, dredging channels, and generally reshuffling the scenery until they had their snug harbors.

San Diego Bay, 10 nautical miles north of the Mexican border, is the home port for several hundred U.S. Navy ships and the site of extensive studies of oceanography and associated meteorology. This bay has natural and man-made harbors providing facilities for approximately 25,000 pleasure boats.

The bay's Camelot-like weather is noted for its moderate temperature, maintained by generally steady, gentle breezes. According to the *U.S. Coast Pilot:* "The winds seldom exceed 25 knots, and shipping is considered safe at all times. Fogs occur occasionally, with heavy fog averaging about 30 days a year. These are mostly night fogs, and seldom occur between the hours of 9 A.M. and 6 P.M."

From San Diego, it is a 55-nautical-mile run along the coast to Newport Harbor. The lovely islands in this harbor are largely man-made creations, the products of imagination and hard work. The passage from Newport Harbor to San Pedro Bay and Los Angeles is about 16 nautical miles.

The eight Channel Islands are situated off the coast in an area from San Diego to northwest of Los Angeles, and they range from 10 to 46 nautical miles from shore. Most of these islands have extremely restrictive landing regulations. However, Avalon Harbor on Santa Catalina Island, which does not require a landing permit, is a popular cruising stopover.

South of Los Angeles, the coastal area bears the brunt of the hot, gusty Santa Ana winds that blast out, with little notice, from the Mojave Desert through the Santa Ana River Canyon and other mountain passes over the waters of the Pacific. These northeasterly or easterly gales have been known to attain velocities in excess of 50 knots. They are most likely to occur during the late fall or winter when a decided barometric pressure difference between a stationary high over the plateau region of Nevada and a low pressure area near this section of the coast appears on weather maps. The effect of a Santa Ana rarely extends more than 40 or 50 nautical miles to sea, but it can cause dangerous seas, particularly in Avalon Harbor, which is exposed to these easterly winds. When the first warning of the approach of a Santa Ana is received by radio on Santa Catalina Island, persons aboard boats are advised to move their boats immediately to the western side of the island.

The *U.S. Coast Pilot* includes the following information concerning the approach of a Santa Ana:

> The barometer is almost useless when its readings are taken alone, for there is little pressure variation, although a gale may spring up and blow for hours. For some hours before a Santa Ana, there is usually a period of good visibility and unusually low humidity. Shortly before its arrival on the coast the Santa Ana may be observed as an approaching dark brown dust cloud. This will often give 10 to 30 minutes warning, and is always one of the positive indications.

Surge must be given special consideration in the harbors near Los Angeles as seiche activity is likely to develop in a manner similar to that in the Great Lakes. The resulting oscillation frequency lasts about an hour, causing rapid changes in the depth of the water and a quick reversal of current. NOAA's publication *Mariners Weather Log* states in an article that in Los Angeles Harbor "large ships have snapped mooring lines and broken pilings because of horizontal movement caused by a seiche."

Coastal fog near the Los Angeles area is usually confined to the nighttime hours, but sometimes dense fog can blanket the harbor for days.

Los Angeles is almost 350 nautical miles south of San Francisco Bay. Cruising between the two locations is difficult because harbors of refuge are scarce along this section of the coast. Offshore boating is too rugged for all but the most experienced. Northbound boats must fight their way into head seas and against prevailing winds as well as deal with the increasing frequency of fog. Forbidding, high, and barren headlands rise from the shore, and offshore rocks can be hazardous for the unwary. One seasoned southern California owner of a power cruiser wrote us that "every year there are several reports of boats in trouble because the skippers just don't know enough about navigation. They go out of the harbor, set the auto pilot, and go barrelling up the coast. They have no idea of how to keep a dead reckoning of their position and are always piling up on shore if the weather closes in."

North/South California Coastal Weather Division The twin promontories of Point Conception and Point Arguello, situated 9 nautical miles apart, are considered to mark the area where the weather conditions prevalent in southern California change to those found in northern California. This area is also referred to as the Fog Line.

Below Point Conception, warm-water eddies are circulated between the coast and the colder California Current, lessening fog-making conditions to the south. This point also tends to hamper the sweep of ocean swells to the south as well, although the seas around both points are large and confused. A note on a *Marine Weather Services Chart* points out that "when a stronger than average Pacific High pressure area is established, northwest winds of 20 to 30 knots frequently occur off Point Conception . . . . These winds, with a duration of 12 to 18 hours, will produce wind waves of 10 to 16 feet."*

North of the points, cold-water upwelling occurs all along the coast, and the incidence of frequent coastal fog is great. "Fogs at Point Arguello are invariably thick," the National Weather Service states, "and this point is recognized by mariners as one of the most dangerous on the coast." The heavy northwesterly gales that often occur off this point are responsible also for its nickname, the Cape Horn of the Pacific.

West Coast sailing friends wrote of their northbound passages as follows:

> We always anchor behind the kelp beds at Cojo, south of the points in order to round them at one or two in the morning, the time when it is supposedly the quietest. While at anchor, the entire crew has been rolled out of their bunks more than once, sometimes even as late as midnight, reportedly the calmest time.

Northern California The entrance to San Francisco Bay must be approached with care. The vast amount of national and international ship traffic here is surpassed in the United States only by the ports of New York and New Orleans. The great numbers of vessels move in and out of the harbor, converging where the main ship channel narrows to about one-half nautical mile at the Golden Gate Bridge, which is suspended more than 60 meters (200 feet) overhead from steep cliffs rising from the water's edge. The speed of the water spilling through this bight has been known to exceed 6.5 knots, creating swirling eddies near the foundations of the bridge piers, which have caused even large ships to be swept off course.

In a description of fog in this area, the *U.S. Coast Pilot* states:

> Golden Gate, the entrance to San Francisco Bay, is a region of frequent fog, and shipwrecks have been numerous there. Often a

---

* 3 to 5 meters.

> sheet of fog forms in early forenoon off the bold headlands on either side of the Golden Gate and becomes more formidable in size as the day wears on. As the temperature rises in the warm inland valleys, a steadily increasing indraft takes place. The fog . . . approaches the shore and enshrouds a good portion or all of San Francisco Bay.

Boating enthusiasts of this area tell us this fog frequently lifts inside the bay and they can pursue their sport inside in good visibility when there is pea soup at the Golden Gate and outside.

The bay area, referred to locally as "the bay," is actually eight interconnecting bays. The two largest are San Francisco Bay, extending from the Golden Gate Bridge southward, and San Pablo Bay adjoining San Francisco Bay to the north. Beyond San Pablo Bay, the Sacramento and San Joaquin Rivers converge to enter this basin by means of the only sea level pass through California's Coast Mountains. The Central Valley, through which these rivers flow prior to their convergence (Sacramento River from the north and San Joaquin River from the south), parallels the coast on the eastern side of the Coast Range for about 645 kilometers (400 miles). The valley is almost completely hemmed in by mountain ranges. When the sun heats the valley, the hot air rises and sucks the coastal air through the pass, pulling it through the Golden Gate to the bay and beyond, thus causing strong westerly winds through the bay area. This wind attains its greatest speed about 4:30 P.M. and is lightest about 6 A.M. One ardent racing skipper tells us that winds of 30 to 40 knots are not at all unusual in San Francisco Bay in the summertime.

The convergence of the Sacramento and San Joaquin Rivers about 64 kilometers (40 statute miles) from the Pacific coast is the beginning of the delta region. This region provides fabulous fishing grounds and protected waters for small boat cruising and gunkholing, all at the back door of the city of San Francisco.

Offshore, from San Francisco north to the Oregon line, boating is primarily carried on by commercial fishermen. Harbors of refuge are very scarce, and some of the areas are often blanketed by the heaviest summer fogs anywhere along this Pacific coast; the sun may not be seen for three or four weeks at a time.

When summer fog forms on the north coast, the temperatures become chilly, with afternoon highs of 10° to 13°C (50° to 55°F). With the wind and dampness, it can be unbelievably uncomfortable, even with foul weather gear and a warm, snug cabin.

When the weather is clear, the occasional recreational boater who makes this passage is treated to a view of the magnificent giant sequoia trees near the coast and in the autumn to the conversational quacking and honking of flock after flock of migrating birds flapping southward down the Pacific Flyway.

Robert A. Baum of the National Weather Service office in Redwood City, Calif., sent us the following information:

> Some note should be made of a not-too-well-known danger (especially to transients) of summer gales from Cape Blanco to Cape Mendocino.
>
> Summer gales occur along the northern California and southern Oregon coasts when the weather is otherwise most pleasant. When the Pacific High builds inland over the Pacific northwest and the thermal trough in the California Central Valley is well established, northerly winds 25 to 40 knots and stronger are not uncommon between Cape Mendocino and Cape Blanco. The combined seas and swell can reach 14 to 20 feet* in as short a time as 6 hours. The frequency of occurrence is probably 10 to 15 times a summer; the duration usually 1 to 3 days.
>
> The strong winds and high seas are dangerous to small boats and create an inconvenience to all vessels. Each year a number of small boats are overpowered by the wind and waves and lost—some of them experienced high seas fishermen. The cold water makes survival of crew members difficult.

Oregon and Washington Cruising along the coasts of these states is a forbidding undertaking. Glaciers have carved rugged basalt faces along the shoreline that withstand the onslaught of winter ocean storms passing through the Aleutian Low. Their profile is softened somewhat by tall majestic pine trees against a background of the volcanic peaks of the Coast Ranges, which often are capped with snow all year long.

The abundance of rocks in the offshore waters here are tokens left by glacial and volcanic activity of the far distant past. Mariners along this portion of the Pacific coast must also contend with the hazards of kelp and driftwood. Evidence of the lumbering industry ashore is often encountered as floating or semisubmerged logs drift past. These tree trunks and logs can be as much as 30 meters (100 feet) long and the ones referred to as sinkers and deadheads have become so watersoaked that they are either submerged just below the surface of the water or one end protrudes above the surface while the rest hangs vertically below.

---

* 4 to 6 meters.

Summer fogs and low stratus clouds are often present along these shores caused by warm moist air that originates in the subtropical Pacific High hundreds of miles away and blows over the cold coastal waters.

Most of the pleasure boating in these states takes place on the protected bays, sounds, and rivers. There are passes in the mountains through which these protected waters empty into the sea. The currents at the seaward outlets and ocean inlets are turbulent. Shifting sandbars make passages difficult and the Pacific swells often break over these bars into the shoal waters close to the shore, especially when a swift ebb tidal current meets heavy seas and swells rolling in from the Pacific Ocean. The hazardous conditions have prompted the U.S. Coast Guard to publish *Bar Guides* for 15 areas. They have also installed bar warning signs at several of the more heavily used inlets. The U.S. Coast Guard chief of boating safety for Oregon and Washington sent us this description of these signs:

> The Bar Warning Signs are diamond-shaped with the words "ROUGH BAR" in black letters and alternating flashing amber lights warning when sea conditions are hazardous. The signs are activated when the seas exceed four feet* in height and are judged to be especially hazardous by the station's Officer-in-Charge.
>
> As stated in the Bar Guides, the fact that the lights are not flashing does not guarantee that sea conditions are favorable. Much of the recreational boating in this area is seasonal with little or no activity during the winter months. Thus many of the stations do not operate the signs from October through April.

The *U.S. Coast Pilot* reports that ebb currents on the northern side of the Columbia River mouth bar attain speeds of six to eight knots and heavy breakers have occurred far inside the entrance. The Columbia River is the boundary between Oregon and Washington.

A 10-nautical-mile-wide waterway, the Strait of Juan de Fuca, cuts into the coastline to separate Vancouver Island in the province of British Columbia, Canada, and the Olympic Peninsula in the state of Washington. The weather at the entrance can be severe, with cross seas, dense fog, and heavy rains. In clear weather, the snow-capped mountain peaks of Washington's Olympic Range are visible.

About 97 kilometers (60 statute miles) inland from the ocean, via this border route, is Puget Sound and its adjacent waters, containing 6500 square kilometers (1,900 square nautical miles) of nearly land-

---

* 1.2 meters.

locked salt water. This picturesque sound serves as a playground for approximately 200,000 pleasure boats. The area includes the jewel-like San Juan Islands, which lie between Vancouver Island and the northwest tip of Washington.

The climate here is dominated by the Aleutian Low in the winter, which limits much of the recreational boating to the seasons when the Pacific High holds sway. Nestled as they are between the Olympic Mountains to seaward and the Cascade Mountain Range to the east, the waters of Puget Sound rarely become seriously rough.

Judy and Doug Zuberbuhler, enthusiastic boating advocates of Puget Sound, tell us that the boating season they observe is generally from May to October because from November through April it *rains* (the italics are theirs). The prevailing winds are usually southerly during the boating season, blowing at an average of 10 to 15 knots. Heavy storms are unlikely, but frequent rain/small-craft-advisory situations do occur. At times like this, it is not unusually dangerous, just wet. Squalls do not blow up ordinarily without four to five hours of visual warning, due, they say, to the sheltered characteristic of Puget Sound. There is very little fog during the boating season, as fog and heavy rain usually are confined to the rest of the year. As far as temperature is concerned, they add, "Even though we had gloriously sunny, warm days on our cruise of the San Juan Islands last summer, we had to wear parkas in the evening and early morning. Puget water temperature is a chilly 40 to 50 degrees Fahrenheit.* There are no insects, however, and this is one of the beauties of Puget Sound sailing."

The San Juan Islands are in the rain shadow of the Olympic Range. Rain falls on the ocean side of the mountains, but it is usually beautifully clear to the northeast of the range where the islands are situated.

Long-distance cruisers can follow the ferry route along the Alaskan Marine Highway from Seattle north through British Columbia to Alaska's capital, Juneau, and be in protected waters almost all of the way.

Alaska's Pacific Coast Pleasure boating that takes place in this northern clime holds little pleasure outside protected waters and the *Marine Weather Services Chart* notes that "small icebergs and bergy bits are present even in August, in the Coastal Waterways." Adventurous souls who take the various protected passages through about 435 nautical miles of British Columbia and the 220 nautical

* 4°C to 10°C.

miles from Dixon Entrance at the Canada-Alaska boundary up the Inside Passage are treated to awe-inspiring scenery. The shores are heavily forested with pines, and snow-capped mountains rise from the water's edge. Glaciers reach icy fingers out through narrow valleys to touch rocky outcrops rimming the land and fjords with gray chiseled walls that seem to rise to touch the sky.

The 220-nautical-mile-long Inside Passage is routed through a wide strip of islands hemmed in by the Gulf of Alaska on one side and the narrow strip of Alaska's panhandle on the other. It ends at Cape Spencer, about 97 kilometers (60 miles) west of Juneau. One entire volume of the eight-book *U.S. Coast Pilot* series is devoted to the locale between Dixon Entrance and Cape Spencer, and included in its valuable information are the following extracts:

> Southeast Alaska is vulnerable to strong winds that often reach gale force . . . . Surrounded by high mountain ranges along the coast, the Gulf of Alaska serves as a catch basin for the often violent storms typical of the northern ocean area . . . . The winds will naturally follow the land contours . . . . There is seldom a month when winds of 60 to 70 knots do not reach some exposed portion of the southeastern coastline.
>
> In summer, nights are often calm; the wind starts up at sunrise, increasing until about 3 P.M., and then gradually moderates after sunset.
>
> Winds passing over the relatively warm ocean area pick up moisture before moving against the mountain terrain, and make this region one of the rainiest in North America.
>
> Light fog prevails over this maritime zone the whole year . . . . Summer fog most often arrives after midnight and may stay until the following noon, or possibly later.
>
> Several large earthquakes have occurred in or near southeastern Alaska during the past 75 years. However, any tsunamis generated have been damaging only near the epicentral area.
>
> Floating logs, deadheads, or sinkers are present through the year in all the inland waters, channels, passes, and inlets in southeastern Alaska and are dangerous to navigation both day and night.

A retired U.S. Coast Guard serviceman who spent two years in southeastern Alaska says that the winters are surprisingly mild in this section of Alaska because a warm current from Japan flows near the coast; there were only two snowfalls during his tour of duty. He has never forgotten the splendor of the steep cliffs lining many of the island passages and the delightful harbors to be found through natural breaks in the walls of rock. He adds, however, that piloting in this area can be very tricky and that "nobody should sail southeastern Alaskan waters without first obtaining and studying the appropriate *U.S. Coast Pilot.*"

One of the most serious hazards to navigation in these waters, he says, is a drifting salmon trap, wrenched from its mooring by a storm or high seas. A trap is a huge framework of rough-hewn logs strapped together to support the fish nets; it can be as large as 18 × 18 meters (60 × 60 feet). Fishermen radio the Coast Guard to report when a trap has broken loose and they instigate a search as soon as possible. The traps are difficult to spot in the water because they are coated with creosote, a black, gunky preservative.

Others who have cruised this section of Alaska rave about the spectacular scenery as well. However, the consensus is that scenery is not the only thing to command attention because wind and current can play strange tricks in the narrow passages and bays. The high Coast Mountains play havoc with wind direction. In some long passages, the wind will be blowing toward the land at one end and will be blowing in the opposite direction at the other end five or six nautical miles away. The currents race around the points of land and through the entrances to inlets. When the current is against the wind, some of the passages are so rough that it is often necessary to wait for the current to change before it is safe to make the trip.

The Townsends met Dick and Kathy Hitchcock in Florida aboard their boat *Alyeska,* with her hail port of Anchorage, Alaska. At that time the Hitchcocks were on their way to the Bahamas with their two small daughters, 3-year-old Tory and 4-month-old Erin, before returning to their home state. Dick spent many hours with Sallie discussing boating in Alaska from a weather point of view, and the rest of this section concerned with Alaska's Pacific coast is a summary of this interview.

Most of the boating in southern Alaska used to be carried on by commercial fishermen, but there are more and more pleasure boats arriving on the scene every year. Many of them power up the Inside Passage from Seattle, a trip any prudent boater can make, as this route is well protected from the open ocean.

Boating in most of the bays along the southern coast of Alaska commences after "breakup," the time when the waters are freed of the ice. Breakup usually happens anytime from April 15 to mid-May. There are really only two seasons in Alaska, winter and summer, because spring and fall rarely last for even two seeks apiece. Although the summer season is short, the days have 20 hours of daylight and persons there really seem to need less sleep during this period. On

June 22, the longest day of the year, nighttime is no darker than twilight.

Weather originates in the Aleutians for all of this coast. If there is a low, "the weather just sits there for one hell of a long time." It is almost always raining or drizzling, and it can be cloudy or rainy for 35 to 40 days in a row. Some years there are only ten sunny days all summer. There is fog more often than not. Pleasure boating is generally suspended during times of very low visibility because navigational aids are almost nonexistent. Near the coast, wind often funnels through the openings between the mountains in gusty blasts. In some places, like Cook Inlet to Anchorage, there is really no pleasure boating at all because of the frequently bad weather and the strong current that sweeps in past the Aleutian Islands. The range of tide at Alaska's Pacific coast is the greatest in the United States, over 10 meters (33 feet) in one section.

Most persons have to travel very far to do their boating. Some drive 500 to 650 kilometers (300 to 400 statute miles) while others fly in their small private planes. There are some 10 000 kilometers (6,200 statute miles) of road in this state but only half of them are paved. The Hitchcocks use their camper to travel to their boat. They have to take everything with them and when they stock up in Anchorage, they buy from the stores that outfit exploration and hunting expeditions. There is no problem keeping things cold; in fact, some of the time Dick remembers using their ice chest to keep things from freezing.

There is no lack of drinking water anywhere, as it is available from any of the springs along the way. Melt from the glaciers, however, is unappetizing, as it is creamy in consistency with a milky blue silt suspended in it, since glaciers melt from the bottom up.

The Hitchcocks always have their fishing equipment along, as fish and crabs are plentiful. Outdoor enthusiasts such as this family can always catch what they need and are grateful for nature's bounty.

Alaskan radio stations give marine broadcasts several times a day. Radio communications to date are good, but VHF-FM communication may be a problem because Alaska is too large and rugged a country for line-of-sight transmission. Citizens band radio for short distances, however, is big up there.

There is no trouble finding helpful and friendly persons. In fact, Dick concludes, "Alaska has to be one of the friendliest places on earth!"

Hawaii Our southernmost state is an isolated archipelago in the north-central portion of the Pacific Ocean. It is composed of many islands and islets totaling a land area of approximately 16 638 square kilometers (6,424 square statute miles). The archipelago stretches from long. 155°W to 178°W and from lat. 19°N to 28°N. It is about 1,800 nautical miles west of San Francisco and south of the Aleutian Islands.

The eight mountainous islands grouped at the southeastern end of the chain are referred to as the High Islands. The state takes its name from Hawaii, known as the Big Island, at the easternmost end of this group. This island, with its area of 10 414 square kilometers (4,021 square statute miles), is almost twice the size of all the others combined. Maui is 22 nautical miles northwest of Hawaii, and the slogan of its residents modestly proclaims, "Maui is best." Three islands are no more than 7 nautical miles distant from Maui: Kahooawe, to the west, is under the jurisdiction of the U.S. Navy and is closed to the public; Lanai, also to the west, is owned by a pineapple company; and Molokai, 19 nautical miles to the northwest, lies between Maui and Oahu.

The state capital of Honolulu is situated on Oahu, the third largest island in the group, and about four-fifths of the population of the Hawaiian Islands live here. Fifty-five nautical miles to the northwest of Oahu is Kauai, the most northerly of the group, and Niihau, 13 nautical miles away from Kaui, is the most westerly of the group. Niihau, primarily populated by native Hawaiians, is owned by a cattle rancher.

A pamphlet entitled *America's Islands,* which is published by NOAA, describes the Hawaiian Island group as follows:

> The 25 or more islands of the Hawaiian Island chain are of volcanic origin and rise from a submerged mountain ridge which stretches in a southeast-to-northwest direction for nearly 2,000 miles in the north Pacific . . . .
>
> The northwestern sector has low coral atolls, sandy islands, reefs and shoals. The peaks reach just to sea level . . . .
>
> In the southeast sector there are eight high volcanic islands . . . . The youngest is Hawaii, which is in a stage of upbuilding and active volcanism; its two 13,000 foot* peaks, Mauna Loa and Mauna Kea, standing 26,000 feet† above the ocean floor, are the highest volcanoes in the world. The other seven islands are extinct volcanoes now in the process of being worn down . . . . These main islands represent 99.9 percent of the dry land area of the entire group, but occupy only a quarter of the length of the Hawaiian chain.

---

* 4000 meters.
† 8000 meters.

Our discussion hereafter is limited to the southeast sector of the islands because this is where the recreational boating action is; only those continuing on with long-distance blue-water cruises go beyond.

The Hawaiian Islands are famed throughout the world for their palm-tree-lined beaches, orchid leis, and luaus. It is a happy place of happy persons, many reflecting in their faces a most successful blending of seafaring ancestors from all over the globe.

The climate is mild throughout the year and the islands are almost always blessed with cool and pleasant breezes, unusual in a region so close to the equator. The delightful temperatures occur because the islands are situated in the planetary wind belt known as the Northeasterly Trades. The *Marine Weather Services Chart* includes the following description of wind in Hawaiian waters:

> August and September in Hawaii are normally warm and dry with persistent trade winds. Over the nearby open sea, these average 13 to 16 knots, and are predominantly from the east-northeast, with directions northeast through east occurring 90 percent of the time in August and about 85 percent of the time in September. However, as summer merges into fall, the trades diminish in frequency and by the end of December occur only about half the time.
>
> The trade winds are ordinarily stronger in the afternoon and lightest in the early morning, the difference being greater in waters close inshore than in those further off.
>
> The rugged and varied terrain of the islands exerts the most pronounced influence on the speed and direction of the wind. Around headlands, in exposed channels, and to the lee of some gorges, passes, and saddles, the trades may be much stronger and gustier than over the open ocean.

On the sheltered or leeward side of the islands where the trade winds are sometimes cut off by high land, these areas are cooled by a sea breeze, which usually develops about mid-morning and lasts until sundown.

The long trade wind swells are considerably less pronounced on the protected sides of the islands and thus are attractive areas for boating. Sailors relying on wind for their power must either sail close to the land to take advantage of the sea breeze or must stay about 10 nautical miles offshore to catch the full benefit of the trades.

The prevailing wind is warm and moist, blowing as it does across warm tropical ocean waters. It is forced upward when it reaches the steep sides of the mountains. As it rises, it cools, and rain results.

There is rain almost every day on the northeast or windward sides of the mountainous islands where rainforests can be found; cactuses grow in the rocky soil and dry air on the other sides.

"Kona" is the Polynesian word for south, and some of the southern or leeward shores on the high islands are referred to as the kona district. Occasionally the trade winds diminish in strength and the wind shifts to blow from the south. This is called the kona wind as it strikes what is usually the leeward shores of the islands. Kona weather, as it is known locally, is likely to occur anytime between October and April. Heavy kona rains are likely to fall on the south sides of the islands at this time and unpleasant humidity is experienced along the northeastern shores.

The volcanic origin of the Hawaiian Islands accounts for the steep drop-off of the ocean floor around each of the high islands. The *U.S. Coast Pilot* notes that the deep water is seldom more than a mile from shore and is usually not far from the coral reefs that fringe much of the coastlines of the islands. The few off-lying dangers to navigation are usually indicated by breakers or by a change in the color of the water. Normally, the deep blue color of the open ocean starts to change to a blue-green between 18 to 27 meters (60 to 90 feet). Bottom features become visible in the clear water at 11 to 13 meters (36 to 42 feet).

NOAA's National Weather Service in Honolulu publishes a comprehensive booklet called *Mariners Weather Guide for Hawaiian Waters and the North Pacific.** It gives the details of all the marine weather forecasts and warnings available for the area. The booklet is published for the benefit of commercial shipping and recreational boaters.

Because of this state's geographical arrangement, weather forecasts in Hawaii are given for the coastal waters within 20 nautical miles of the shorelines, inter-island channels, and waters within 1,000 nautical miles of Honolulu.

The occasional hurricane and tsunami warnings issued are broadcast by Hawaii radio stations. The islands are rarely affected by tropical cyclones because, according to the *U.S. Coast Pilot,* they "lie on the extremities of both the western north Pacific typhoon area and the eastern north Pacific hurricane area."

Fluttering proudly on the bow of a gleaming white cruiser moored next to the Townsends on one of their Intracoastal Waterway

---

* Order from: NOAA, National Weather Service, Honolulu, Hawaii 96820.

trips was the burgee of the Hawaii Yacht Club, recognizable because of the state's name emblazoned across its face. It must have taken Sallie all of thirty seconds to strike up an acquaintance with its owner in order to query him about his yacht club's boating water.

Captain Hoover, the owner and a retired airline pilot, was fortunate to have spent much of his time in Hawaii and did a lot of boating while he was there. He pointed out that "cruising between the islands is not for amateurs!" The wind blasts through the channels and the seas are strong once you are out of the lee of the islands. He confirmed that almost all of the boating takes place on the downwind side of the islands where there is plenty of wind, but the seas do not build up because the waters are shielded by the mountainous terrain. However, a few miles out from this protection, the seas are once again very high.

Gerry Ward, a former resident of the islands and present racing competitor of the Townsends, says that September through November, late April, and all of May are the ideal times for small boating. These times fall between the months when huge swells triggered by distant winter storms break on Hawaiian shores, and those when the summer trades, reinforced by the Pacific High, are especially robust.

The most startling thing about sailing here, he adds, is the crystal-clear visibility that is taken for granted by the islanders. It comes as a great shock, therefore, when this incredible visibility is lowered to a mile or so in vog, as it is called locally. Vog occurs during kona weather when a volcano acts up on the island of Hawaii. The southeasterly kona wind spreads ash clouds from the island of Hawaii along the whole string of High Islands to enshroud them in a foglike vapor; when the trade winds hold sway, any ash clouds are blown to sea away from the islands. The volcanic culprit is usually the Kilauea Crater of Mauna Loa.

Pleasure boating in the islands is increasing steadily. Experienced blue-water skippers sometimes bring their own boats across the Pacific from the mainland. We understand, however, that there are charter boats available at a few locations, but usually charters come with captains. Understandably, owners are loath to turn over their boats to newcomers even when they are experienced elsewhere.

From December into April, as mentioned before, huge swells from far-off northern storms reach the islands; some of these swells are said to originate as far away as Siberia, 1,300 nautical miles away, and from the southwest and south, 2,200 nautical miles away. The winter surfing competitions on Oahu are followed with great interest on the mainland. Television and moviers viewers gasp at the size of the huge curling combers ridden by the tiny figures balancing on their boards.

The nobility of Hawaii were riding these huge rollers on boards long before the first traders and missionaries set foot on the islands. Thus, the young persons of today who have made surfing a way of life in our island state can quite correctly refer to it as a "sport of kings" in this realm of enchantment.

APPENDIX A

The Metric System: Meteorological Units

The metric system is widely used internationally and, with the Metric Act of 1975, the U.S. Metric Board and the National Bureau of Standards began the transition process necessary for the adoption of this system in the United States. The concept of the simple standards of the metric system was proposed in France by Gabriel Mouton in 1670. It was not until 1875 that an International Bureau of Weights and Measurements was formed in Sèvres, France, where it is still located. This bureau is governed by an international committee responsible for standardizing the Système International d'Unites (International System of Units), commonly referred to as SI Units.

The Metric Act of 1975 did not specify a time limit for the SI Units changeover in this country. NOAA assures marine interests that "ample notice will be given the public and dual units will be used during the transition period."

The key to the metric system is the set of prefixes, divisible by 10, which can be applied to basic units of measurement.

SI PREFIXES **(Partial list)**

Multiples & Submultiples	*Prefixes*	*Symbols*	*Relationship to Base Unit*
1000	kilo	k	1000 times the unit
100	hecto	h	100 times the unit
10	deka	da	10 times the unit
1	— — — — — — —		Single base unit
0.1	deci	d	1/10 of a unit
0.01	centi	c	1/100 of a unit
0.001	milli	m	1/1000 of a unit

The prefix followed by the base unit becomes one word. Using a meter as the base unit, the following terms, for example, may be interchanged to express an equal measurement:

1000 meters = 1 kilometer

1/100 meter or 0.01 meter = 1 centimeter

1/1000 meter or 0.001 meter = 1 millimeter

When speed is expressed, a slash mark is followed by the abbreviation for the time interval, such as:

per hour = /h
per minute = /m
per second = /s

Thus,

meters per second = m/s
kilometers per hour = km/h

Among the units which are not SI but will be maintained because they are so widely in use, according to the National Bureau of Standards, are: second, minute, hour, and day indicating time; second, minute, and degree indicating latitude and longitude; and degree indicating compass direction of wind.

Figure 23 illustrates the metric measurements adopted by the National Weather Service in 1975 for reporting weather conditions to the general public as well as their counterpart U.S. customary measurements prior to this time.

Following are some of the rules for writing metric numbers as set forth by the international committee for the SI Units:

	Example
A zero always precedes a decimal when the number is less than one	0.05
Spaces are used instead of commas to separate sets of three digits to the left of the decimal point	1 234 500
Spaces are used to separate sets of three digits to the right of the decimal point	0.123 45
When four digits are not used in a column of numbers, they may be written with or without a space	1234 or 1 234

UNITS USED TO MEASURE WEATHER CONDITIONS - METRIC AND U.S. CUSTOMARY

WEATHER	METRIC	U.S. CUSTOMARY
Atmospheric Pressure:	kiloPascals (kPa) (1 kPa = 10 mb = 0.30 in)	millibars (mb) inches of mercury (in)
Temperature:	degrees Celsius (°C) $°C = \frac{5 \times (°F - 32)}{9}$	degrees Fahrenheit (°F)
Dew Point:	Same as temperature	Same as temperature
Wind Direction:	Same as U.S. Customary	Reported: degrees compass (°) Forecast: N, NE, E, SE, etc.
Wind Speed:	kilometers per hour (km/h) (1 km/h = 0.54 kn = 0.62 mph)	knots (kn) miles per hour (mph)
Precipitation Amount:	millimeters (mm) (1 mm = 0.04 in)	inches (in)
Snow Depth: (Water Equivalent)	centimeters (cm) (1 cm = 0.39 in)	inches (in)
Ice Thickness:	centimeters (cm) (1 cm - 0.39 in)	inches (in)
Visibility:	kilometers (km) (1 km = 0.54 nm = 0.62 mi)	nautical miles (nm) statute miles (mi)
Heights: clouds shore waves river stage tides	meters (m) (1 m = 3.28 ft)	feet (ft)

Fig. 23. UNITS USED TO MEASURE WEATHER CONDITIONS: Metric and U.S. Customary

APPENDIX B

Temperature Conversions: Celsius—Fahrenheit & Wind Chill

Celsius (C), previously referred to as centigrade, and Fahrenheit (F) are different units for the measurement in degrees of temperature based on the same standards: the freezing point of water at sea level (0°C and 32°F) and the boiling point of water (100°C and 212°F). The following procedures may be used to convert one to the other:

Fractions	(9/5 = 1.8)	*Decimals*
$°F = \frac{9 \times °C}{5} + 32$		°F = °C times 1.8 plus 32
$°C = \frac{5 \times (°F - 32)}{9}$		°C = °F minus 32 divided by 1.8

Temperature degrees in the text are converted to the nearest whole degree.

It might be helpful to use units of ten on the Celsius scale as an aid to remember the equivalent Fahrenheit degrees and the type of weather generally associated with these temperatures for winter and summer days.

Degrees		***Weather***	
C	*F*	*Winter*	*Summer*
−30°	−22°	Bitter cold	
−20°	− 4°	Very cold	
−10°	+14°	Cold	
0°	32°	Freezing level	
+10°	50°	Mild	Cold
20°	68°		Mild
30°	86°		Hot
40°	104°		Heat wave

TEMPERATURE CONVERSIONS: CELSIUS TO FAHRENHEIT

°C: 50, 40, 30, 20, 10, 0, -10, -20, -30

°F: 120, 110, 100, 90, 80, 70, 60, 50, 40, 30*, 20, 10, 0, -10, -20, -30

°C	°F	°C	°F	°C	°F
-35	-31.0	- 5	23.0	25	77.0
-34	-29.2	- 4	24.8	26	78.8
-33	-27.4	- 3	26.6	27	80.6
-32	-25.6	- 2	28.4	28	82.4
-31	-23.8	- 1	30.2	29	84.2
-30	-22.0	0	32.0	30	86.0
-29	-20.2	+ 1	33.8	31	87.8
-28	-18.4	2	35.6	32	89.6
-27	-16.6	3	37.4	33	91.4
-26	-14.8	4	39.2	34	93.2
-25	-13.0	5	41.0	35	95.0
-24	-11.2	6	42.8	36	96.8
-23	- 9.4	7	44.6	37	98.6
-22	- 7.6	8	46.4	38	100.4
-21	- 5.8	9	48.2	39	102.2
-20	- 4.0	10	50.0	40	104.0
-19	- 2.2	11	51.8	41	105.8
-18	- 0.4	12	53.6	42	107.6
-17	+ 1.4	13	55.4	43	109.4
-16	3.2	14	57.2	44	111.2
-15	5.0	15	59.0	45	113.0
-14	6.8	16	60.8	46	114.8
-13	8.6	17	62.6	47	116.6
-12	10.4	18	64.4	48	118.4
-11	12.2	19	66.2	49	120.2
-10	14.0	20	68.0	50	122.0
- 9	15.8	21	69.8		
- 8	17.6	22	71.6		
- 7	19.4	23	73.4		
- 6	21.2	24	75.2		

* Freezing point of water at sea level

Fig. 24. TEMPERATURE CONVERSIONS: Celsius to Fahrenheit

TEMPERATURE CONVERSIONS: FAHRENHEIT TO CELSIUS

°F	°C	°F	°C	°F	°C	°F	°C	°F	°C
-30	-34.4	0	-17.8	30	- 1.1	60	15.6	90	32.2
-29	-33.9	+ 1	-17.2	31	- 0.6	61	16.1	91	32.8
-28	-33.3	2	-16.7	32	0.0	62	16.7	92	33.3
-27	-32.8	3	-16.1	33	+ 0.6	63	17.2	93	33.9
-26	-32.2	4	-15.6	34	1.1	64	17.8	94	34.4
-25	-31.7	5	-15.0	35	1.7	65	18.3	95	35.0
-24	-31.1	6	-14.4	36	2.2	66	18.9	96	35.6
-23	-30.6	7	-13.9	37	2.8	67	19.4	97	36.1
-22	-30.0	8	-13.3	38	3.3	68	20.0	98	36.7
-21	-29.4	9	-12.8	39	3.9	69	20.6	99	37.2
-20	-28.9	10	-12.2	40	4.4	70	21.1	100	37.8
-19	-28.3	11	-11.7	41	5.0	71	21.7	101	38.3
-18	-27.8	12	-11.1	42	5.6	72	22.2	102	38.9
-17	-27.2	13	-10.6	43	6.1	73	22.8	103	39.4
-16	-26.7	14	-10.0	44	6.7	74	23.3	104	40.0
-15	-26.1	15	- 9.4	45	7.2	75	23.9	105	40.6
-14	-25.6	16	- 8.9	46	7.8	76	24.4	106	41.1
-13	-25.0	17	- 8.3	47	8.3	77	25.0	107	41.7
-12	-24.4	18	- 7.8	48	8.9	78	25.6	108	42.2
-11	-23.9	19	- 7.2	49	9.4	79	26.1	109	42.8
-10	-23.3	20	- 6.7	50	10.0	80	26.7	110	43.3
- 9	-22.8	21	- 6.1	51	10.6	81	27.2	111	43.9
- 8	-22.2	22	- 5.6	52	11.1	82	27.8	112	44.4
- 7	-21.7	23	- 5.0	53	11.7	83	28.3	113	45.0
- 6	-21.1	24	- 4.4	54	12.2	84	28.9	114	45.6
- 5	-20.6	25	- 3.9	55	12.8	85	29.4	115	46.1
- 4	-20.0	26	- 3.3	56	13.3	86	30.0	116	46.7
- 3	-19.4	27	- 2.8	57	13.9	87	30.6	117	47.2
- 2	-18.9	28	- 2.2	58	14.4	88	31.1	118	47.8
- 1	-18.3	29	- 1.7	59	15.0	89	31.7	119	48.3
								120	48.9

Fig. 25. TEMPERATURE CONVERSIONS: Fahrenheit to Celsius

WIND CHILL EQUIVALENT TEMPERATURE

According to a bulletin issued by the Environmental Data Service for the National Weather Service, "The wind chill index or equivalent temperature is based upon a neutral skin temperature of 33°C.* With physical exertion, the body heat production rises, perspiration begins, and heat is removed from the body by vaporization. The body also loses heat through conduction to cold surfaces with which it is in contact and in breathing cold air that results in the loss of heat from the lungs. The index, therefore, does not take into account all possible losses of the body. It does, however, give a good measure of the convective cooling that is the major source of body heat loss."

The following Wind Chill Equivalent Temperature table has been adopted by the International System of Units (SI). We added the knot figures for your convenience. To convert to Fahrenheit, use the conversion table for temperature (Fig. 24).

---

* 91.4°F.

WIND CHILL EQUIVALENT TEMPERATURE IN DEGREES CELSIUS

- at speeds in knots and kilometers per hour -

		WIND SPEED										
	(knots)	3	5	11	16	22	27	32	38	43	49	54
	(km/h)	6	10	20	30	40	50	60	70	80	90	100
	20	20	18	16	14	13	13	12	12	12	12	12
D	16	16	14	11	9	7	7	6	6	5	5	5
R	12	12	9	5	3	-1	-0	-0	-1	-1	-1	-1
Y	8	8	5	0	-3	-5	-6	-7	-7	-8	-8	-8
	4	4	0	-5	-8	-11	-12	-13	-14	-14	-14	-14
B												
U	0	0	-4	-10	-14	-17	-18	-19	-20	-21	-21	-21
L	-4	-4	-8	-15	-20	-23	-25	-26	-27	-27	-27	-27
B	-8	-8	-13	-21	-25	-29	-31	-32	-33	-34	-34	-34
	-12	-12	-17	-26	-31	-35	-37	-39	-40	-40	-40	-40
T	-16	-16	-22	-31	-37	-41	-43	-45	-46	-47	-47	-47
E												
M	-20	-20	-26	-36	-43	-47	-49	-51	-52	-53	-53	-53
P	-24	-24	-31	-42	-48	-53	-56	-58	-59	-60	-60	-60
E	-28	-28	-35	-47	-54	-59	-62	-64	-65	-66	-66	-66
R	-32	-32	-40	-52	-60	-65	-68	-70	-72	-73	-73	-73
A	-36	-36	-44	-57	-65	-71	-74	-77	-78	-79	-79	-79
T												
U	-40	-40	-49	-63	-71	-77	-80	-83	-85	-86	-86	-86
R	-44	-44	-53	-68	-77	-83	-87	-89	-91	-92	-92	-92
E	-48	-48	-58	-73	-82	-89	-93	-96	-98	-99	-99	-99
	-52	-52	-62	-78	-88	-95	-99	-102	-104	-105	-105	-105
(°C)	-56	-56	-67	-84	-94	-101	-105	-109	-111	-112	-112	-112
	-60	-60	-71	-89	-99	-107	-112	-115	-117	-118	-118	-118

Fig. 26. WIND CHILL EQUIVALENT TEMPERATURE

APPENDIX C

LINEAR DISTANCE CONVERSIONS: METERS - FATHOMS - FEET

Meters	Fathoms	Feet
1	0.546 81	3.280 84
2	1.093 61	6.561 68
3	1.640 42	9.842 52
4	2.187 23	13.123 36
5	2.734 03	16.404 20
6	3.280 84	19.685 04
7	3.827 65	22.965 88
8	4.374 45	26.246 72
9	4.921 26	29.527 56

Fathoms	Meters	Feet
1	1.828 80	6
2	3.657 60	12
3	5.486 40	18
4	7.315 20	24
5	9.144 00	30
6	10.972 80	36
7	12.801 60	42
8	14.630 40	48
9	16.459 20	54

Feet	Meters	Fathoms
1	0.304 80	0.166 67
2	0.609 60	0.333 33
3	0.914 40	0.500 00
4	1.219 20	0.666 67
5	1.524 00	0.833 33
6	1.828 80	1.000 00
7	2.133 60	1.166 67
8	2.438 40	1.333 33
9	2.743 20	1.500 00

Fig. 27. LINEAR DISTANCE CONVERSIONS: Meters–Fathoms–Feet *To find the equivalents for numbers that are multiples of 10, simply move the decimal points one place to the right in the appropriate figures in the table for each zero added. Example:*

300 meters = 164 fathoms or 984 feet

To use the table for numbers that are not multiples of 10, such as 148 meters, procede as follows:

100 meters = 328.084 feet
40 meters = 131.234 feet
8 meters = 26.247 feet

148 meters = 485.565 or *486 feet* (answer)

You may prefer to multiply the single-unit conversion times the number, especially if you use a calculator. In this case:

148 × 3.280 84 = 485.564 32 or 486

Linear Distance Conversions: Meters–Fathoms–Feet

The changeover to meters affects reports in cloud, wave, and river heights; precipitation amounts; and heights and depths given in *Light Lists, Tide Tables,* and nautical charts.

"The Defense Mapping Agency Hydrographic Center," according to a 1977 announcement in a Coast Guard *Notice to Mariners,* "is continuing the program to gradually convert the depths and heights on nautical charts and in publications to the metric system . . . . The fact that the depths are shown in meters (usually in meters and decimeters* to 20 meters) is boldly stated in the chart title and in magenta-colored type in the outer chart borders."

To help you remember the relationship of the metric system to the U.S. customary, some approximations of conversions from the meters–fathoms–feet tables are listed below as well as some approximations that are relevant to the conversions but do not appear in the tables:

1 meter = ½ fathom (0.546 81 fm)
= 1 1/10 yards (1.093 61 yd)
= 3¼ feet (3.280 84 ft)
= 39 inches (39.370 08 in)

1 centimeter	= 2/5 inch (0.393 70 in)
1 millimeter	= 1/25 inch (0.039 37 in)
1 fathom	= 2 meters (1.828 80 m)
1 yard	= 9/10 meter (0.914 40 m)
1 foot	= 3/10 meter (0.304 80 m)
	= 30 centimeters (30.480 00 cm)
1 inch	= 2½ centimeters (2.540 00 cm)
	= 25 millimeters (25.400 00 mm)

* 1 decimeter = 0.1 meter.

APPENDIX D

Linear Distance Conversions: Kilometers–Nautical Miles–Statute Miles

In the metric system, kilometers generally replace statute or land miles. Nautical miles will be maintained for mariners for some time to come since they are an internationally accepted standard for measuring distances over water. According to a communiqué issued by the National Bureau of Standards in 1975, the use of nautical miles is accepted for a limited time, subject to future review.

In the nineteenth century in France, 10 000 000 *metres* (meters) were considered to be the length of the quadrant of the earth from the equator to the north pole with geodetic measurements made from Dunkerque, France, to Mont Jouy, Spain. The meter has since been redefined and currently is related to a wavelength of light and not an arc of the earth.

Today, latitude degrees are used to measure north-south distances from the equator and nautical miles can be measured from latitude scales on navigational charts. The relationship of latitude to nautical miles and the earth's surface is outlined below:

1 minute (1′) of latitude = 1 nautical mile
1 degree (1°) of latitude = 60 nautical miles (There are 60 minutes of latitude in one degree)
0° latitude = equator
90° North (90°N) latitude = North Geographical Pole
90° South (90°S) latitude = South Geographical Pole

Some handy approximations that can be used with the kilometers–nautical miles–statute miles tables are listed below:

1 kilometer = ½ nautical mile (0.539 96 nm)
= 3/5 statute mile (0.621 37 mi)
1 nautical mile = 1850 meters (1852 is the exact measurement adopted for the international nautical mile)
= 2 kilometers (1.852 00 km)
= 1 1/7 statute miles (1.150 78 mi)
1 statute mile = 1600 meters (1 609.34 m)
= 1 3/5 kilometers (1.609 34 km)
= ⅞ nautical miles (0.868 98 nm)

Fig. 28. LINEAR DISTANCE CONVERSIONS: Kilometers—Nautical miles—Statute miles. SPEED CONVERSIONS: Kilometers per hour—Knots—Statute miles per hour *To find the equivalents for numbers that are multiples of 10, simply move the decimal points one place to the right in the appropriate figures in the table for each zero added. Example:*

300 kilometers = 162 nautical miles or 186 statute miles

To use the table for numbers that are not multiples of 10, such as 148 kilometers, procede as follows:

100 kilometers = 53.996 nautical miles
40 kilometers = 21.598 nautical miles
8 kilometers = 4.320 nautical miles

148 kilometers = 79.914 or *80 nautical miles* (answer)

You may prefer to multiply the single-unit conversion times the number, especially if you use a calculator. In this case:

$$148 \times 0.539\,96 = 79.914\,08 \text{ or } \underline{\underline{80}}$$

LINEAR DISTANCE CONVERSIONS: KILOMETERS - NAUTICAL MILES - STATUTE MILES

SPEED CONVERSIONS: KILOMETERS PER HOUR - KNOTS - STATUTE MILES PER HOUR

KILOMETERS or km/h | NAUTICAL MILES or KNOTS | STATUTE MILES or mph

Kilometers or km/h	Nautical Miles or Knots	Statute Miles or Statute mph
1	0.539 96	0.621 37
2	1.079 91	1.242 74
3	1.619 87	1.864 11
4	2.159 83	2.485 48
5	2.699 78	3.106 86
6	3.239 74	3.728 23
7	3.779 70	4.349 60
8	4.319 65	4.970 97
9	4.859 61	5.592 34

Nautical Miles or Knots	Kilometers or km/h	Statute Miles or Statute mph
1	1.852 00	1.150 78
2	3.704 00	2.301 56
3	5.556 00	3.452 34
4	7.408 00	4.603 12
5	9.260 00	5.753 90
6	11.112 00	6.904 68
7	12.964 00	8.055 46
8	14.816 00	9.206 24
9	16.668 00	10.357 02

Statute Miles or Statute mph	Kilometers or km/h	Nautical Miles or Knots
1	1.609 34	0.868 98
2	3.218 69	1.737 95
3	4.828 03	2.606 93
4	6.437 38	3.475 90
5	8.046 72	4.344 88
6	9.656 06	5.213 86
7	11.265 41	6.082 83
8	12.874 75	6.951 81
9	14.484 10	7.820 79

APPENDIX E

Speed Conversions: Kilometers per Hour–Knots–Statute Miles per Hour

The tables for speed units are found in Figure 28.

Speed conversions needed to determine the speed of storm movements across the country, wind speed, boat speed, or the like, must be made in compatible units:

kilometers per hour	—kilometers
knots	—nautical miles
	(1 knot = 1 nautical mile per hour)
statute miles per hour	—statute miles

Some approximations from the speed-conversions table follow along with some approximations that are relevant to the conversions but do not appear in the tables:

1 kilometer per hour	= ½ knot (0.539 96 kn)
	= $^{3}/_{5}$ statute mile per hour (0.621 37 mph)
	= ¼ meter per second (0.277 78 m/s)
1 knot	= 2 kilometers per hour (1.852 00 km/h)
	= $1^{1}/_{7}$ statute miles per hour (1.150 78 mph)
	= ½ meter per second (0.514 44 m/s)
1 statute mile per hour	= $1^{3}/_{5}$ kilometers per hour (1.609 34 km/h)
	= ⅞ knot (0.868 98 kn)
	= $^{2}/_{5}$ meter per second (0.447 04 m/s)
1 meter per second	= $3^{3}/_{5}$ kilometers per hour (3.600 00 km/h)
	= 2 knots (1.943 84 kn)
	= 2¼ statute miles per hour (2.236 94 mph)

APPENDIX F

Atmospheric Pressure Conversions: KiloPascals–Millibars–Inches of Mercury

The metric unit adopted for atmospheric air pressure is the kiloPascal. It is equal to 10 millibars and approximately ⅓ inch of mercury. Millibars have been used for many years for recording atmospheric pressure and, in the United States, inches of mercury have been the customary unit used in weather reports intended for the public.

A generally accepted standard measurement for atmospheric pressure is based on a pressure figure considered to be normal at sea level at a latitude of 45 degrees with an air temperature of 15°C (59°F). Following is this standard atmosphere figure for each of the three units of measurement and the conversions of one unit to another:

KiloPascals (kPa): standard atmospheric pressure = 101.33 kPa
1 kiloPascal = 10 millibars
1 kiloPascal = 0.30 inch of mercury
Millibars (mb): standard atmospheric pressure = 1013.25 mb
1 millibar = 0.10 kiloPascal
1 millibar = 0.03 inch of mercury
Inches of mercury (in): standard atmospheric pressure = 29.92 in
1 inch of mercury = 3.39 kiloPascals
1 inch of mercury = 33.86 millibars

ATMOSPHERIC PRESSURE CONVERSIONS: KILOPASCALS - MILLIBARS - INCHES OF MERCURY

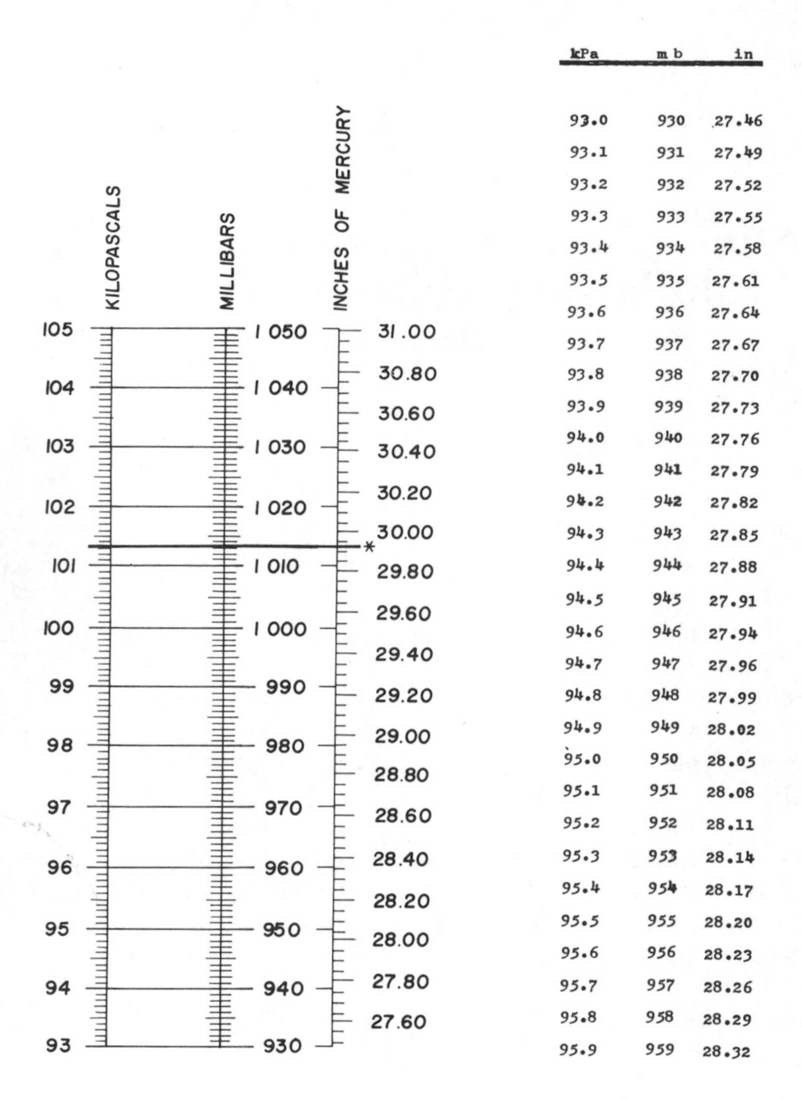

kPa	mb	in
93.0	930	27.46
93.1	931	27.49
93.2	932	27.52
93.3	933	27.55
93.4	934	27.58
93.5	935	27.61
93.6	936	27.64
93.7	937	27.67
93.8	938	27.70
93.9	939	27.73
94.0	940	27.76
94.1	941	27.79
94.2	942	27.82
94.3	943	27.85
94.4	944	27.88
94.5	945	27.91
94.6	946	27.94
94.7	947	27.96
94.8	948	27.99
94.9	949	28.02
95.0	950	28.05
95.1	951	28.08
95.2	952	28.11
95.3	953	28.14
95.4	954	28.17
95.5	955	28.20
95.6	956	28.23
95.7	957	28.26
95.8	958	28.29
95.9	959	28.32

*Standard atmosphere at sea level at 45°N latitude and 15°C (59°F) temperature:

kPa	mb	in
101.325	1013.250	29.921 26

Fig. 29. ATMOSPHERIC PRESSURE CONVERSIONS: KiloPascals–Millibars–Inches of Mercury

ATMOSPHERIC PRESSURE CONVERSIONS: KILOPASCALS - MILLIBARS - INCHES OF MERCURY

kPa	mb	in	kPa	mb	in	kPa	mb	in
96.0	960	28.35	99.0	990	29.23	102.0	1020	30.12
96.1	961	28.38	99.1	991	29.26	102.1	1021	30.15
96.2	962	28.41	99.2	992	29.29	102.2	1022	30.18
96.3	963	28.44	99.3	993	29.32	102.3	1023	30.21
96.4	964	28.47	99.4	994	29.35	102.4	1024	30.24
96.5	965	28.50	99.5	995	29.38	102.5	1025	30.27
96.6	966	28.53	99.6	996	29.41	102.6	1026	30.30
96.7	967	28.56	99.7	997	29.44	102.7	1027	30.33
96.8	968	28.59	99.8	998	29.47	102.8	1028	30.36
96.9	969	28.61	99.9	999	29.50	102.9	1029	30.39
97.0	970	28.64	100.0	1000	29.53	103.0	1030	30.42
97.1	971	28.67	100.1	1001	29.56	103.1	1031	30.45
97.2	972	28.70	100.2	1002	29.59	103.2	1032	30.47
97.3	973	28.73	100.3	1003	29.62	103.3	1033	30.50
97.4	974	28.76	100.4	1004	29.65	103.4	1034	30.53
97.5	975	28.79	100.5	1005	29.68	103.5	1035	30.56
97.6	976	28.82	100.6	1006	29.71	103.6	1036	30.59
97.7	977	28.85	100.7	1007	29.74	103.7	1037	30.62
97.8	978	28.88	100.8	1008	29.77	103.8	1038	30.65
97.9	979	28.91	100.9	1009	29.80	103.9	1039	30.68
98.0	980	28.94	101.0	1010	29.83	104.0	1040	30.71
98.1	981	28.97	101.1	1011	29.86	104.1	1041	30.74
98.2	982	29.00	101.2	1012	29.89	104.2	1042	30.77
98.3	983	29.03	101.3	1013	29.92	104.3	1043	30.80
98.4	984	29.06	101.4	1014	29.94	104.4	1044	30.83
98.5	985	29.09	101.5	1015	29.97	104.5	1045	30.86
98.6	986	29.12	101.6	1016	30.00	104.6	1046	30.89
98.7	987	29.15	101.7	1017	30.03	104.7	1047	30.92
98.8	988	29.18	101.8	1018	30.06	104.8	1048	30.95
98.9	989	29.21	101.9	1019	30.09	104.9	1049	30.98
						105.0	1050	31.01

*Standard atmosphere at sea level at 45°N latitude and 15°C (59°F) temperature:

kPa	mb	in
101.325	1013.250	29.921 26

Index

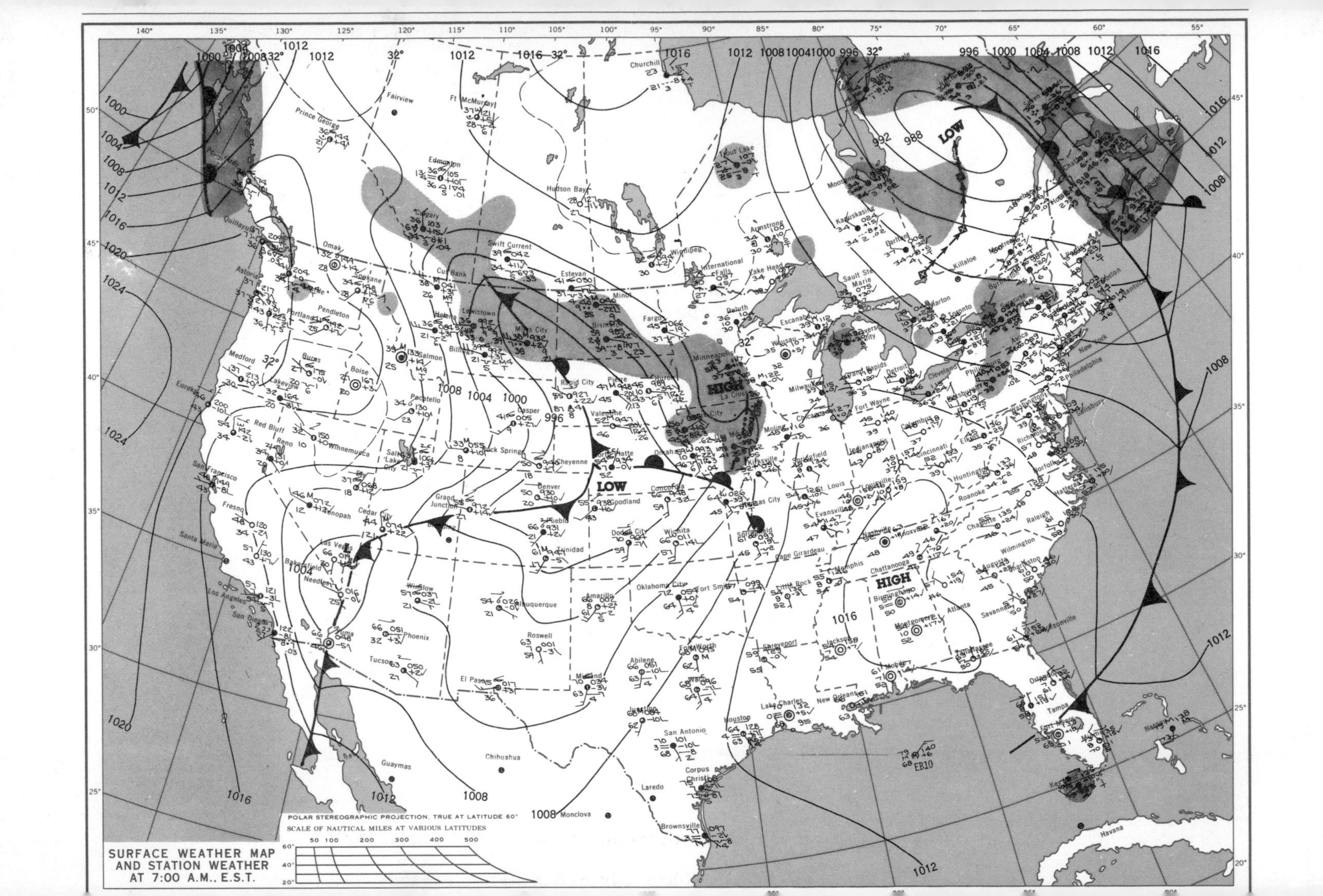
SURFACE WEATHER MAP
AND STATION WEATHER
AT 7:00 A.M., E.S.T.
POLAR STEREOGRAPHIC PROJECTION, TRUE AT LATITUDE 60°
SCALE OF NAUTICAL MILES AT VARIOUS LATITUDES
50 100 200 300 400 500
HIGH
LOW
HIGH
LOW
HIGH
LOW